THE GREAT QUAKE DEBATE

THE GREAT
QUAKE DEBATE

*The Crusader, the Skeptic,
and the Rise of Modern Seismology*

SUSAN HOUGH

UNIVERSITY OF WASHINGTON PRESS
Seattle

Design by Jarrod Taylor

Composed in Miller Text, typeface designed by Matthew Carter

24   23   22   21   20   5   4   3   2   1

Printed and bound in the United States of America

UNIVERSITY OF WASHINGTON PRESS

uwapress.uw.edu

LIBRARY OF CONGRESS CATALOGING-IN-PUBLICATION DATA

Names: Hough, Susan Elizabeth, 1961–, author.

Title: The great quake debate : the crusader, the skeptic, and the rise of modern seismology / Susan Hough.

Description: Seattle : University of Washington Press, [2020] | Includes bibliographical references and index.

Identifiers: LCCN 2019045528 (print) | LCCN 2019045529 (ebook) | ISBN 9780295747361 (hardcover) | ISBN 9780295747378 (ebook)

Subjects: LCSH: Willis, Bailey, 1857–1949. | Hill, Robert Thomas, 1858–1941. | Seismology—California, Southern—History—20th century. | Geologists—United States—Biography. | Earthquake prediction—California, Southern—History—20th century. | Earthquakes—California, Southern—History—20th century.

Classification: LCC QE535.2.U6 H682 2020 (print) | LCC QE535.2.U6 (ebook) | DDC 551.2209794/909041—dc23

LC record available at https://lccn.loc.gov/2019045528

LC ebook record available at https://lccn.loc.gov/2019045529

Author photo by Gino DeGrandis, with permission of the Seismological Society of America

The paper used in this publication is acid free and meets the minimum requirements of American National Standard for Information Sciences—Permanence of Paper for Printed Library Materials, ANSI Z39.48–1984.∞

Los Angeles is situated a favored locality sufficiently distant from the active faults upon which great earthquakes are generated. . . . So far as destructive shocks are concerned Los Angeles has not acquired the habit, and if we may judge by the record, is not likely to acquire the habit.

—Andrew Lawson,
February 23, 1927

CONTENTS

PREFACE

I set out to write the story of a scientific debate that played out in the early twentieth century, splashing outside of sheltered academic waters and into the news: does the Los Angeles area face significant earthquake hazard? It might be hard to imagine the answer was ever in doubt. Sometimes something is so well known that we forget that somebody figured it out in the first place. And people outside of the scientific community might not realize the extent to which the figuring-out process can be contentious. Science can be a messy business, if not a contact sport. The process by which science moves forward has as much to do with people and personality—persuasion and push-back—as with instruments or data. When science touches on matters of societal relevance, outside forces sometimes come into play. Scientists tend to view people and personality as beside the point, a distraction from the things that really matter (i.e., science). But as much as it is a science story, this book is a people story, the story of two individuals who came to embody the debate. You probably don't know their names, let alone their science, much less who they were as people. I barely knew them myself when I began this project, and in the course of spending many hours with the writings they and others left behind, I realize that much of what I thought I knew turns out to have been incomplete, if not outright wrong. May I have the pleasure of introducing Robert Thomas Hill and Bailey Willis, the costars of the great quake debate.

THE GREAT QUAKE DEBATE

PROLOGUE
Setting the Stage

> Gentlemen, I want this picture to start with
> an earthquake and work up to a climax.
> —*Samuel Goldwyn*

Studies of instrumentally recorded modern earthquakes are alike; historical earthquakes are historical in their own way. Without question, the earthquake that struck California on the morning of January 9, 1857, was both portentous and historical in its own way. Hours before dawn that morning, California began to stir, initially with gentle rumblings reported at a small handful of locations including Santa Cruz and San Francisco. More people, from Santa Barbara to San Francisco, reported shaking shortly after first light that morning, and again close to sunrise. While it remains impossible to pinpoint the locations of these initial shocks with only scattered accounts of felt shaking, the distribution of reports points to a region that remains remote today, not far from the dusty stretch of State Highway 46 near the town of Cholame where James Dean met his untimely death a century later.

Modern scientists can barely characterize the early-morning foreshocks, the largest of which is estimated to have been magnitude 5.5–6.0.

Of the earthquake that began at 8:24 a.m. on January 9, 1857—its timing established by a handful of especially precise eyewitness reports—we know quite a bit more. This temblor rocked the young state of California nearly stem to stern, generating perceptible shaking that extended northward beyond Sacramento, east to Mission Las Vegas, and south beyond the Mexican border. The 1857 earthquake struck at a pivotal time in California's history. Had this temblor occurred a hundred years earlier, it would likely not have been felt by people who kept written records. Scientists today would know about it only from the signature it left behind etched in the landscape.

But by the early nineteenth century, Catholic priests of the Franciscan order had built twenty-one missions along El Camino Real from San Diego to San Francisco, and by 1857, California had entered the Union and had started to establish itself as an American population hub. Unlike the native tribes who had lived there for thousands of years, the new settlers had a tradition of written record-keeping. By 1856, the population of the new state had reached about 500,000, including about 65,000 Native Americans. Fueled by the gold rush of 1849, by 1860 San Francisco had joined the ranks of the top twenty largest US cities, with a population of more than 50,000. Beyond the Bay Area, only a single other California city had cracked the top one hundred largest US cities as of 1860: Sacramento, with a population of almost 14,000. To the south, the city of Los Angeles existed by 1857, but the discovery of oil that propelled it to its current behemoth status was still a half century away. In the mid-nineteenth century it remained a remote outpost. Census data reveal a population of 4,385 persons in Los Angeles in 1860, just 3 percent of the state population. Other settlements had sprung up elsewhere in California, including mining towns on both sides of the Sierra Nevada and a small Army outpost at Fort Tejon, at the northern end of the corridor later known by generations of California drivers as the Grapevine.

In Central California south of Cholame, settlements in the Carrizo Plain remained essentially nonexistent in 1857, but an occasional cowboy wandered through with sheep or cattle. One cowboy, known to history as Mr. Bell, reported felt shaking before daybreak, strong enough to cause his cattle to stampede. Mr. Bell's account, relayed and transcribed circa 1905, poses something of a puzzle. Once it became light enough, Bell reported that he and his helpers started to search for the cattle. Although he described seeing considerable dust at the foot of the Temblor Range, which he assumed to have been generated by the cattle, he did not describe feeling violent shaking—or any shaking for that matter—as he searched. Some scientists would later conclude that years afterward, Bell simply misremembered the time of the initial shock, that it—and the cattle stampede—must have been later. But 8:24 a.m. was well after daybreak that January morning, and therein lies the puzzle. Bell didn't just report a time, he reported an entire narrative, the details of which make no sense if the cattle stampede happened well after sunrise. If Bell's recollections were correct, the apparent intimation is that he and his men did not notice shaking from the 8:24 a.m. main shock as they rode through the Carrizo Plain on horseback.

Geological evidence now tells us that the two sides of the San Andreas Fault, which scientists now know to be the primary plate-boundary fault separating the North American and Pacific Plates, slipped past each other some 10 feet (3–4 m) in this earthquake. It is hard to imagine that the shaking from this event could have escaped the attentions of people who were nearly on top of the fault, even on horseback. Could the fault have unzipped in a fast but relatively smooth way, such that shaking immediately along the fault remained relatively subdued? It is an intriguing possibility, that shaking directly atop this great earthquake might not have been severe enough to be noticed by a person on horseback. But how much stock can one put in a single after-the-fact recollection of one individual? *Every historical earthquake is historical in its own way.* The information available to modern science regarding this and other important historical earthquakes is limited, not necessarily reliable, and sometimes enigmatic. Sometimes, seemingly puzzling accounts can tell us something important about an earthquake. Sometimes puzzling accounts are simply wrong.

Still, riddles notwithstanding, it makes all the difference for our understanding of the 1857 earthquake that there were not only witnesses, but witnesses who documented their experiences in writing. News traveled more slowly then than now, and the sparse population kept the effects of the temblor limited, but in time reports from around the state found their way into newspapers. The following day, the *San Francisco Herald* reported that "the earth was shaken to its centre by the throes which seem to have become a part of the peculiarities of our state." The modern reader detects a note of whimsy in reports of less-than-catastrophic, but still sometimes dramatic, effects in San Francisco. "On Market Street a small frame house was shaken from its foundation and fell a distance of five or six feet—though it is probable a high wind would have done as much for it just at that time." And on Stewart Street, "the shock produced an emptying of the houses of their human contents—upwards of fifty persons ran frightened and confused into the street." Elsewhere throughout the city, boards toppled in lumberyards, and light objects and dishes fell in the southern part of the city in particular: "all the accompaniments of a second-class earthquake." One notes here that, a full half century before 1906, San Franciscans had—or thought they had—some idea what a first-class earthquake feels like.

In our social media age, photographs of an earthquake in California would be beamed around the world in less time than it takes for seismic waves to travel the same distance. A hundred and fifty-odd years ago, it took about a month for East Coast newspapers to gather and relay reports to their readers. On February 15, 1857, the *New York Herald* published eyewitness accounts from around California. In many parts of the state, reports echoed those from San Jose: "The movement was undulating and slow . . . and produced a sickening sensation precisely as one feels when upon the rocking wave." The tenor of the accounts changed moving south. In Santa Barbara, a letter writer described "the most terrible earthquake experienced for the last twenty-six years in California," with "several houses injured." Some of the most dramatic reports in this article came from the small Army outpost at Fort Tejon. Following a light, barely perceptible shock around six in the morning, the shock reported at half-past eight that morning lasted "from three to five minutes," leaving devastation in its wake. "Nearly all of the buildings in the vicinity were seriously injured by the falling of chimneys, plastering, and walls. Several adobe buildings in the course of erection and nearly completed were almost totally destroyed." A medical doctor at the fort reported having been "violently thrown from his feet." At a government mill 20 miles (33 km) from the fort, mules were thrown down, branches of trees were broken, and "large oaks fell to the ground." Thanks to natural selection, large trees withstand strong winds quite well, which gives them a healthy measure of earthquake resilience; it is not easy to break a tree.

At the time of the earthquake, a young woman named Augusta lived with her family on a farm near Lytle Creek, some 15 miles (24 km) northwest of San Bernardino. She, too, painted a vivid and violent scene: "At the same instant I saw my parents and sisters clinging to large trees, whose branches lashed the ground, birds flew irregularly through the air shrieking, horses screamed, cattle fell bellowing on their knees, and even the domestic feathered tribe were filled with consternation. Voices of all creatures, the rattling of household articles, the cracking of boards, the falling of bricks, the splashing of water in the wells, the falling of rocks in the mountains and the artillery-like voice of the earthquake, and even that awful sound of the earth rending open—all at once, all within a few seconds, with the skies darkened and the earth rising and falling beneath the feet—were the work of an earthquake." Piecing together geological evidence and historical accounts, scientists later established that the 1857 earthquake began near Cholame and blasted south-southeast on the San Andreas Fault through the arid Carrizo Plain. Toward the end of the Carrizo Plain the rupture continued through a kink in the fault, barreling on for another 90 miles (145 km) or so before running out of steam somewhere close to Augusta's family farm.

The earliest scientific reports refer to the 1857 "earthquake in California." In 1923, geologist Bailey Willis published an article in the *Bulletin of the Seismological Society of America* titled "Earthquake Risk in California." In this article he appears to have bestowed the name that stuck, the Fort Tejon earthquake. By (informal) convention, earthquakes today are generally named after the town or geographical feature closest to the epicenter, which we locate using waves recorded by modern seismometers. At the time the Fort Tejon earthquake struck, scientists did not entirely understand what an earthquake *is*, which is to say, they did not understand the physical process that plays out in the earth that generates felt shaking. The very name *earthquake*, from the Greek *seism*, referred to the shaking of the earth, not the physical process that caused the shaking. Before the end of the nineteenth century, speculations about the nature of the physical process were generally off base. The association between earthquakes and movement along two sides of a fault first

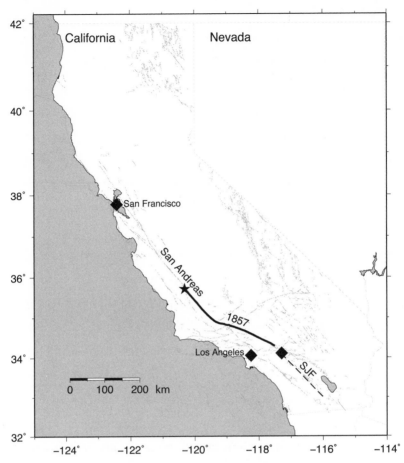

San Andreas Fault and other faults in California, including the San Jacinto Fault (SJF, dashed line); estimated extent of fault rupture in the 1857 Fort Tejon earthquake (solid line), as well as locations of San Francisco, San Bernardino, and Los Angeles (diamonds).

emerged following the 1891 Mino-Owari earthquake in Japan. Faulting had been observed previously, but was taken to be among the secondary effects of an unknown deeper process. In the aftermath of the great 1906 San Francisco earthquake, scientists' understanding of the earthquake source process deepened with the development of elastic rebound theory, which explains how stress builds along a fault—as a consequence of forces that remained unknown at the time—and gets released in earthquakes.

Today the word *epicenter* is familiar not only to scientists but has entered the popular lexicon, to the extent that the minor-league Rancho Cucamonga Quakes play at a ballfield known as the Epicenter, with mascots Tremor and Quake. Robert Mallet, an Irish engineer who became interested in seismology, coined the word *epicenter*—derived from the Latin *epicentrum*, which in turn was derived from the Greek *epickentros*—in the mid-nineteenth century. At a time when scientists did not yet understand the physical process that causes the earth to quake, *epicenter* referred to the point on a map where the disturbance had been centered. As scientists developed a better understanding of earthquakes over the following decades, they also worked out the basic wave types that earthquakes create. In the early twentieth century, scientists first developed methods to use the observed timing of these waves arriving at different sites to determine an earthquake's location in the crust. The word *hypocenter* was introduced, by which we mean "epicenter plus depth." In this context, *location* refers to the specific point in the crust where a fault first starts to break, where an earthquake nucleates. Every earthquake, big or small, begins at a point, where a tiny patch of a fault begins to give way and the initial movement snowballs. Scientists today understand the snowballing process better than we do the more basic questions: Why does an earthquake nucleate in the first place? And why do most earthquakes stay small but some grow large?

By the time Bailey Willis wrote his 1923 article, scientists understood that earthquakes begin at a point and involve movement across two sides of a fault. They moreover understood that a large earthquake can involve movement along a long swath of a fault, about 300 miles (480 km) in the case of the 1906 San Francisco earthquake. Decades later, when the 1994 Northridge, California, earthquake jolted the greater Los Angeles area, seismologists scrutinized the data carefully before concluding that the hypocenter was, just barely, within the town limits of Northridge rather than the neighboring Reseda and naming it accordingly. Looking back at older earthquakes, with little to no data from modern seismometers, scientists like Willis generally named earthquakes after what newspapers used to call the seat of the disturbance. And so did Willis immortalize the 1857 temblor as the great Fort Tejon earthquake, the name by which scientists know it today. A map included in his 1923 paper identified locations from Yuma, in Arizona just beyond the southeast tip of California, to well into the Central Valley, where Willis reported that effects had been described as "violent," with "moderate shaking noted as far north as Sacramento." In Willis's words, it was "probably one of the most violent earthquakes known to have occurred in California." The best modern magnitude estimate for the earthquake is 7.7, based on careful analysis of detailed geological offsets along the fault.

According to Google Scholar, which tracks how scientific publications are cited by later papers, Willis's 1923 paper has been lost in the sands of time, with only a handful of citations by later scholarly articles. It is, however, not only an interesting read but a paper that was scarcely inconsequential at the time. The article begins with section 1, The Point of View: "In this and succeeding articles earthquake risk in California is to be handled from the point of view of insurance. It may be ignored or ridiculed, prayed against or fled from, or examined and guarded against." Willis went on to say, "The only safeguard against the forces of nature, whether they be lightning strikes or earth tremors, is *understanding*, by virtue of which we may be forearmed because forewarned." In this article, Willis laid out with no small measure of eloquence a key driving force behind modern earthquake science: we study earthquakes and the shaking they generate so we can reduce earthquake risk.

Bailey Willis emerges as one of two pivotal figures in a heated debate that played out over the quarter century following the 1906 San Francisco earthquake. The 1906 temblor was comparable in size to the 1857 earthquake, but ruptured the northern segment of the San Andreas Fault, from San Juan Bautista northward to beyond Cape Mendocino. One might think that this earthquake established beyond all shadow of doubt that, at a minimum, Northern California faces high earthquake hazard. In fact, even that seemingly obvious point remained debated to at least some extent. By 1900 the population of the state was close to 1.5 million; a decade later it had nearly doubled, to just under 2.4 million people. More people meant more businesses, and more businesses meant more business interests and city boosters. News also traveled faster in 1906 than it had a half century earlier. By April 19, 1906, the day following the great earthquake, headlines around the country screamed: "MANY BURIED ALIVE IN RUINS AND BURNED. NO HOPE REMAINS OF SAVING SAN FRANCISCO." Beyond front-page stories that blared in 48-point headlines, follow-up articles informed readers that the temblor had been no fluke, with headlines like "CALIFORNIA A CENTER OF EARTHQUAKE ZONE."

Headlines notwithstanding, what Gladys Hansen later dubbed the denial of disaster began before the embers of San Francisco cooled. Literally as San Francisco still burned on April 19, 1906, California Governor George C. Pardee informed then governor of Massachusetts Curtis Guild Jr. that the devastation in San Francisco had been due overwhelmingly to the firestorms that followed, not the earthquake itself. By the end of that month, the Real Estate Board of San Francisco had resolved to "speak hereafter of the disaster as 'the great fire,' rather than 'the great earthquake.'" The disinformation campaign was not entirely disinformation: to a large extent, San Francisco had been reduced to rubble and ash by the fires, not the shaking itself. But the earthquake that city boosters downplayed had, of course, been a momentous event in its own right. And some took the disinformation campaign further, going well beyond what science could say at the time (or now) to say that the recent occurrence of the temblor meant that San Francisco was safe from destructive earthquakes for centuries to come.

Whether the campaigns of San Francisco city boosters were entirely successful in erasing the association between earthquakes and their corner of the world can be debated. Damaging earthquakes make for splashy news. The 1906 earthquake itself garnered banner headlines in newspapers around the country, but even modestly damaging earthquakes in Northern California had been front-page news since the earliest days of the gold rush. The 1906 earthquake added an emphatic exclamation point to an association that had been established earlier by the 1868 Hayward earthquake and smaller temblors that struck the San Francisco Bay area during the nineteenth century. By 1906, however, few people had experienced or even heard about the 1857 temblor, and nobody remembered that, back in 1770, the San Gabriel valley north of Los Angeles had experienced so many strong earthquakes that Father Junípero Serra dubbed it the Valle de los Temblores. In Southern California, the 1906 earthquake sparked a more nuanced, more complex, and far more intense debate.

Among the scientific community, the 1906 earthquake cemented the understanding that both Southern and Northern California face significant earthquake hazard. For starters, it led directly to the mapping of almost the entire San Andreas Fault, although, as an interesting historical footnote, geologists of the day believed that long-term motion on all faults, including the San Andreas, was primarily vertical. The earthquake they had just witnessed, which clearly ripped roads and fences sideways, was explained away as a fluke. Still, they might not have understood the fault, but they were able to follow it, not only to the southern terminus of the 1906 break near San Juan Bautista, but as far as San Bernardino. South of the Carrizo Plain, where signs of the 1857 temblor could still be found a half century later, the San Andreas hugs the northern edge of the San Gabriel Mountains, part of the so-called Transverse Ranges. While they are not the Himalayas or even the Sierra Nevada, the San Gabriel Mountains rise nearly 2 miles (more than 3,000 m) at their highest point, a big block of rock separating the San Andreas Fault from the flatlands to their immediate south, now home to the sprawling Los Angeles metropolitan area—or the Southland, as it is sometimes called today. Thus, whereas the San Andreas menaces San Francisco more directly, running almost right under the south support of the Golden Gate Bridge, it is

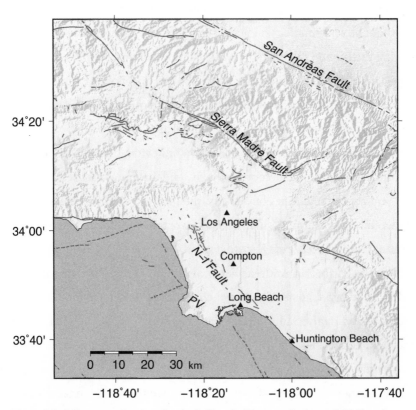

Mapped fault lines near Los Angeles, including the Newport-Inglewood (N-I) and Palos Verdes (PV) Faults. The San Andreas Fault runs along the northern flank of the San Gabriel Mountains, north of Los Angeles.

Prologue

farther removed from Los Angeles. Thus, the great quake debate centered on the question, Are earthquakes also a real and present danger in Los Angeles?

The debate about earthquake hazard in the Southland—which effectively began with one earthquake in 1906, picked up steam in the mid-1910s, and ended with another earthquake in 1933—hinged only partly on the San Andreas, which geologists recognized as an active major fault a full half century before it would be understood to be the primary boundary between two of the earth's massive tectonic plates. The debate also, however, involved the shatter pattern of faults that had been mapped within the Los Angeles basin itself. To what extent are these faults active? What kind of shaking could earthquakes on these faults potentially produce? The debate involved a complex interplay of scientific and societal issues, as well as many individual scientists. But at the heart of the story are the intertwined stories of two individuals, the two geologists who eventually personified the debate. In the conventional telling of this story, these men played well-defined roles: one the flawed hero, the other the villain. Bailey Willis was one of these men; the other was his contemporary, Robert Thomas Hill. This story about Southern California thus begins on the other side of the country, with the stories of two remarkable men and how, while they were alike in many ways, they came to square off so publicly against one another in the great quake debate.

CHAPTER 1

BAILEY WILLIS

> Your wanderer must go to the limits of his tether.
> —*Bailey Willis*

Scientists do not tend to be famous. By virtue of introducing the magnitude scale that bears his name, Charles Richter remains the one earthquake scientist whose name and contributions are widely known among the general public. Before the publication of Richter's landmark paper introducing the magnitude scale in 1935, scientists had no way to measure the intrinsic size of an earthquake. Among the community of earthquake scientists, of course, more names are known, the pioneers whose seminal contributions pushed earthquake science forward to today's modern field of scientific inquiry. We recognize, among others, Robert Mallet for furthering our understanding of earthquake waves, geologist G. K. Gilbert for pioneering insights into the nature of faults and the earthquake cycle, Harry Reid for the development of elastic rebound theory, and Inge Lehmann (seismology's most famous founding mother) for her discovery of the earth's solid inner core. Among the modern scientific community, Bailey Willis's name remains relatively obscure. In fact, although he became passionately interested in earthquake hazard late in his career, and wrote a book describing faults and other geological features, his scientific research never focused on earthquakes.

Who was Bailey Willis? Ironically, or perhaps fittingly, Willis came into the world just a few months after the earthquake he would later name. Born on May 31, 1857, in the town of Cornwall, New York, Willis chose his parents wisely. His father, Nathaniel Parker Willis, was a well-known poet and journalist—a literary celebrity—with family roots in the New World tracing back to seventeenth-century Boston. His mother, Cornelia Grinnell Willis, known to family and friends as Nellie, was Nathaniel's second wife, whom he married after his first wife died in childbirth. Cornelia was an accomplished individual, about whom somewhat more is known than her role as a woman behind a great man—or, in her case, great men. Cornelia Grinnell Willis, born in 1825, has been described, and not only by her son, as a force to be reckoned with: an early women's rights advocate, a founding member of the first women's club in New York, and a dedicated abolitionist. In 1846 she married Nathanial Willis, who brought with him a young daughter from his first marriage. Cornelia went on to have five children, the last of whom, a daughter, died at birth in 1860, leaving Bailey to grow up as the youngest child in the family. The next youngest sibling, Edith, was nearly four years older than him.

In the later words of her son, Cornelia Willis "managed house, garden, and stables, directing numbers of Negro servants of whom, as I now know, only a few were permanent." Cornelia Willis was, Bailey wrote, "antislavery, and true to her principles." During and before Bailey's childhood, the Cornwall house provided safe haven to fugitive slaves until they could be safely dispatched to Canada. Among the permanent paid staff in the Willis household was a woman, Harriet Jacobs, who had been hired in 1842 by Nathaniel's first wife. Nathaniel's first wife had apparently never known that Harriet was a fugitive from North Carolina. Reportedly treated well as an employee of the Willis family, in 1850 Harriet was hounded by the man who had bought her rights from the slave owner from whom she had escaped. Cornelia learned of Harriet's background and bought her freedom in 1852 for the then-not-insubstantial sum of three hundred dollars. This gesture might have gone unnoticed by history, except that Harriet went on to notable accomplishments of her own. Living with the Willis family in the early 1850s, Harriet began working on a manuscript recounting her life as a slave. Published as a book in 1861, *Incidents in the Life of a Slave Girl* made Harriet Jacobs a celebrity in her own right. Together with her daughter Louisa, she later ran a boarding-house in Cambridge, Massachusetts.

Cornelia and Nathaniel built their home on an expansive piece of property in Cornwall, New York, an estate they named Idlewild, five years before Bailey was born. The house sat atop a promontory that afforded a grand view of the Hudson River below. Idlewild gave young Bailey room to roam. A big black Newfoundland dog named Caesar featured prominently in his earliest childhood memories, as well as a golden setter unimaginatively named Goldie. Together Bailey and his canine companions explored every inch of Idlewild, including a brook, ledges, and steep cliffs. His was a free-range childhood, but always within the confines of the estate and without other children to play with. The dogs, he wrote later, "were [my] only playmates."

To her son Bailey, Cornelia was more than a loving and devoted parent; she was his teacher, his mentor, his guiding compass. His early years were spent "free like Mowgli, but sheltered and taught by his Mother's love." Bailey did not learn to read until he was seven years old, but his acquaintance with literature began years earlier. His mother read to him frequently, favoring tales of adventure like that of Sir Samuel Baker's

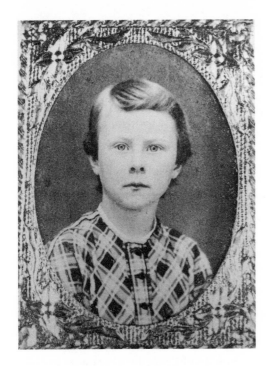

Bailey Willis as a young boy (undated). (Huntington Library photograph used with permission)

exploration for the sources of the Nile. His own four-legged companion Caesar was a veteran of Elisha Kane's expeditions in the Arctic, stories on which Bailey was weaned from his earliest days. "From Mamma," he later wrote, he "learned what brave men can face and overcome," adding, "Not till years later did [I] realize her own courage had far surpassed that of any man, however heroic."

He later credited his father with having passed along his love of nature, courage, and "reverence for true sentiment." But his own reverence he reserved for his mother. "To my Mother," he later wrote, "I can never do justice. She was my inspiration to do my best and the ruling spirit of my every worthy act." Older siblings may have spared Bailey the not uncommon fate of the indulged only child, but it would be understandable if a mother doted on her second-youngest child after losing the child who should have been the baby of the family.

Willis wrote later that the Civil War "cast a shadow" on Idlewild in 1861, but in 1864 a much deeper pall was cast by "the shadow of lingering death." In that year, Nathaniel fell ill, suffering violent epileptic seizures, his overall health quickly spiraling downward. Before this time, Cornwall had been "a home of open-handed hospitality," a welcoming mecca to "the choicest spirits of the literary world of America and England, as well as travelers of distinction." With the onset of Nathaniel's illness, which quickly left him bedridden, Idlewild closed in on itself. Cornelia refused to allow her husband to be cared for by others, insisting that he would remain at home. She stayed true to this resolution "through three agonizing years," even as financial pressures weighed increasingly heavily on the family with the loss of Nathaniel's income. To help meet expenses, Cornelia opened a boarding school for young women, not only teaching Latin and English literature herself, but also taking on the entire burden of management and care for her charges. While Bailey continued to "[go] his individual way" much of the time, with little if any interaction with the students, his mother always found time for him. She "called him her 'Sunshine,' and he, perhaps, though unconscious of her great need, tried to be sunshine to her. She was the whole world to him."

Nathaniel did not loom large in his son's later memories. Nothing Bailey wrote later suggests that his father was a major presence in his life even before he became ill, although he was told that his father used to take him up on his saddle at times and joke about the contrast between the spritely boy and his own six-foot-tall bearded self. When his mother told her children in a broken voice on the morning of January 20, 1867, "You have no Father more, dear children," it was not a deep blow to the young boy. Bailey later recalled having only been "troubled because she was." For his own part, he "had no sense of loss. His Father had never been one of his companions." After he had taken ill, Nathaniel was to his son "the man lying propped up on his pillows, whom he had seen occasionally," someone "to be sorry for; that was all." For practical if not emotional reasons, the senior Willis's death in 1867 marked the beginning of the second chapter of Bailey's formative years. Idlewild was sold; years later Bailey learned that the lawyer who handled the sale stole the proceeds, leaving Cornelia dependent on her father, Joseph Grinnell, a longtime Massachusetts congressman.

The Willis family moved to Cambridge, Massachusetts, where Willis's older brother Grinnell, known in the family as Nelt, was "rowing himself through Harvard." Of note for Bailey, "Grandfather Joseph had definite ideas" about his young grandson's education. Joseph Grinnell continued to provide for his daughter and for Bailey's education. In 1870 the family, including sister Edith, moved to London. In 1871, Cornelia made arrangements for Bailey to stay with a family who lived in a castle, Schloss Lahneck, in the German village of Bingen am Rhein, and to attend the local village school. For the first time in his fourteen years, Bailey "was separated from his mother and left among strangers." He "sobbed as they drove up to the castle." Bailey got past the initial trauma and soon came to relish the setting, the "impassive strength" of the castle itself, "the freedom of roaming in the vineyards and forest." It was not unlike Idlewild, but boasted an actual medieval castle, with an actual (and formidable) castle tower. It was "in the tower that [his] harvest of fancies was richest." He played there often, and at one point, "driven by [his] curiosity to dare the scare, [he] obtained permission to sleep there." With candle in hand, frightened half to death, he climbed the worn circular steps to the room where Goethe had written his poem "Geistergruss" (Greetings to the Ghosts). Bailey wrote, "Then I would dive into my little bed and lie shivering, but happy." His initial apprehensions about the arrangement quickly faded. By early December he wrote to his mother, "It is all jolly here."

Willis was less impressed by the school itself. "It was not much of a school," he wrote later, "but Mother probably thought it might serve as an introduction to the less familiar surroundings and discipline of real German teaching." Bailey enjoyed his years in the castle, "running wild like a colt in the fields." His mother, however, grew concerned about his education, or lack thereof. She selected a private school in Frankfurt, "said to have excellent discipline." Indeed, the school provided a more rigorous academic setting, as well as other things. In Willis's later words, "There I first met with unkindness and learned to fight." Along with two other American students, Bailey soon had run-ins with "two Englishmen, who were older and much stronger individually." They were also not especially kind. "According to their traditions, it was their right as upper classmen to order us to fag for them and they were by no means moderate or gentle in their ways." A self-described "nice little boy," he took lessons from one of the other American boys at the school, a "born fighter" who "trained us in our parts." The young Willis might have been a nice little boy, but the son of Cornelia Willis was no pushover. Together, the three young Yankee boys "eventually taught [their] tormentors that Americans would not submit to British tyranny." Their planned military campaign left them "knocked about," but "the bullies got black eyes and ever afterwards let us alone." Willis described the experience as a most valuable lesson, his "initiation in independence."

After a year in Frankfurt he landed in an *oberrealschule* near Stuttgart, where, "under severe but salutary discipline," he embarked on studies that prepared students for careers other than law or medicine. Although his living arrangements were not ideal during these years, he enjoyed the school and his classmates there. He remained in Germany until the summer of 1874, when he left to attend the Columbia University School of Mines, the first mining and metallurgy school in the United

States. Cornelia also left England to reestablish a home in New York so that she could be with her son during his time at Columbia. Although Bailey felt more German than American by this time, he again thrived in his new environment, earning a degree in mechanical engineering in 1878 and, a year later, in civil engineering. The man who had been born the same year as the great Fort Tejon earthquake finished his university education in 1879, the same year that the United States Geological Survey— "the Survey," as it is sometimes known—was created, primarily to map the country's mineral resources.

Following his graduation, Willis did not have his sights on a particular position, but was eager to find any job. One of Willis's university professors recommended him to the Survey's first director, Clarence King. Sitting outside of King's office in 1879, waiting to learn of his first assignment, "a tall man breezed past me" and entered the office unannounced. "I was astonished. Who was this who could thus unceremoniously approach the Director of the Geological Survey?" "Hello, King!" the man said. "Hello, Pumpelly," King replied. The two proceeded to discuss the plan for sampling the iron ores of the United States. "You will want young men," King eventually said. "There is Willis, you can have him."

And so did Willis embark on the first steps of his scientific career, working with economic geologist Raphael Pumpelly. Although Willis initially continued to work for the Survey, the Northern Pacific Railway Company paid his salary during later work he undertook under Pumpelly's supervision. In a life with no shortage of good fortune, landing with Pumpelly as the first supervisor of his professional career was among the many happy notes. It is possible that more than mere chance brought the men together in King's office: for a time, Pumpelly had lived in the Cambridge, Massachusetts, boardinghouse run by Louisa and Harriet Jacobs. Willis later described Pumpelly as "a distinguished mining engineer and explorer of great ability and charming personality. A flowing red beard marked him in any company, and his genial manner and ready conversation immediately put the listener at ease." By the time he and the young Willis crossed paths, he had dodged the gendarmerie with Corsican mountaineers, escaped raiding Apaches in Arizona, taught miners in Japan how to fire blasts, traveled by pony sleigh across Siberia and Russia, and "faced great dangers in China, where he was known as the 'Red Devil'." Pumpelly was not only a "delightful raconteur" with many fine stories to tell, but also a supervisor who "gave generously to advance his assistants in the humanities as well as in geology, with never a thought of self, but constantly with the desire to develop knowledge."

And so in 1879, Willis began his professional career under Pumpelly's direction, tasked with examining iron ore deposits with an eye toward future economic exploitation. Willis's job description involved, basically, traveling to remote hinterlands, from southern Appalachia to northern Minnesota to Washington State, estimating coal and iron reserves. It cannot have been an easy life for a young man, let alone one born into genteel country estate life. Bailey Willis was moreover slight of stature, energetic and wiry, weighing in at only 125 pounds (56 kg) at 5 feet 9 inches (1.75 m) tall. By early adulthood, however, the slightly built country boy had learned to stand his ground and how to fight, successfully, when he had to fight. Bailey Willis not only met the challenges head on, drawing from deep reserves of both physical and intellectual energy, but relished the experiences. In the forests of Michigan he learned the rudiments of surveying by pacing and compass. In the south, he found ore banks from which the Confederates made iron "with the aid of the gray-bearded mountaineers and moonshiners whom [he] encountered on a solitary ride of 600 miles through the valleys and mountains of the South." A less enterprising and confident individual might have fared less well on this journey.

Dispatched later to chase down reports of iron ores north of Lake Superior, Willis embarked on a long trip with Ojibway Indians that "began in birchbark canoes and ended on snowshoes." Along with his Native American companions and two woodsmen, he reached the Vermilion Iron Range in November 1890, just as winter set in. (Anyone familiar with the upper Midwest might wonder about the timing of the expedition.) Donning snowshoes, Willis continued across the ridge, taking magnetic observations and studying the rocks. "The winds howled and rose to a blizzard," he wrote, "and the temperature went to 20 degrees below zero." The Ojibways did not, however, "let blizzards interfere with their friendly intentions." One morning, as he and his companion snowshoed, they heard a gunshot and saw two Native Americans leaving the camp. "Like suspicious whites," he later wrote, "we assumed that our blankets and grub might be leaving with them. We hurried down to find not only everything in order as we had left it but a fine catch of white fish for our supper."

After three weeks Willis had surveyed the range to his satisfaction, and the party set out for Duluth, 90 miles (145 km) south through a thick forest with no trails. "We were seven, four Indian packers and three white," he wrote. "Color," said the son of Cornelia Willis, "made no difference. We slept side by side and if one flopped we all flopped." One of the woodsmen, Willis continued, "was most popular. He had the habit of smoking a pipeful at 2:00 a.m. and would then build up the fire." The journey, not surprisingly, proved arduous. In Willis's words, both of the woodsmen "froze their toes." The best man among the group, Willis wrote, was an "Ojibway half-breed" named Paul. "He carried the heaviest pack, and in the morning would break trail ahead of me and almost run a hundred yards just to show how it might be done." Paul eventually confessed to having been scared at the start of the trip, pointing to his companions' lack of experience. But he added with a smile, "No scared now, not any more."

Willis later reunited with Pumpelly in Portland, Oregon. At this point, Pumpelly was the director of the Northern Transcontinental Survey, the organization established by the president of the Northern Pacific Railway to discover and map resources along the road then being built between North Dakota and the Pacific Coast. Pumpelly had chosen Willis to explore for coal in the Northwest: Oregon, Washington, and Idaho. Coal reserves were known to exist on Puget Sound, but at the edge of a heavily forested wilderness. Beneath the nearly impenetrable cover of forestland, rich coal lands were believed to exist. In Willis's words, he was "ill-prepared" for the job at hand. The prospect of searching through dense forest to find coal did not scare him. "Others had found it, and I could." The organizational challenges, however, gave Willis pause: the need to organize crews, build bridges, and lead teams of men far from a base of supplies. He might have turned to an "Old Gang" of railroad officials for advice, but they were opposed to the president of the Northern Pacific Railway and, by extension, anyone who worked for him. Willis followed the orders he had been given to avoid the Old Gang. In his words, "They set traps, baited with bribes and girls; but I was much too innocent to be caught."

The young geologist began to make his way, a "tenderfoot in [a] tough society." "I neither smoked, drank, nor swore," he wrote. "I had no training in the management of men." But he brought to the task the abiding sense of fairness and responsibility that Cornelia Willis had instilled in him, combined with tough edges honed first in the school yard in Frankfurt and later during blizzards in the Iron Ranges. Tasked with assessing the value of a remote coal mine, Willis found a spot in the woods to build a log cabin as a base, leading a dozen woodsmen to the shantytown of Wilkeson, about 25 miles (40 km) southeast of Tacoma. "They were men I had picked up in the saloons of Tacoma," Willis wrote, "much against my will, for the environment revolted me." His reservations notwithstanding, Willis came to appreciate the skill of the axmen, including two "choppers" who one day cut down a tree that clipped Willis's head when it fell. Willis initially wrote off the mishap as a consequence of his own carelessness, but redheaded foreman Paddy Jim cursed the two choppers and dismissed them on the spot. The rest of the crew did not like the dismissal, nor did Willis. But confronted with what he saw as an issue of authority, Willis upheld his foreman's decision and then told Paddy Jim that he, too, was dismissed. "I won't have men cursed in my camp," Willis told him. Jim accepted the decision, and the rest of the crew went back to work without further incident, joking within Willis's earshot about their "boy boss."

From the established base, Willis met with Skookum Smith, who had a coal mine to sell. Willis's assignment was to determine whether or not it was worth the sale price of $150,000 plus five cents' royalty per ton of coal mined. "I had no idea how to value a coal mine," he later wrote, "but orders were orders." He first met Smith, whom he described as "six feet three and broad in proportion," in contrast to his own more modest physical dimensions. Introducing himself as the company representative and a mining geologist, Smith asked, "Does your mother know you are out, my boy?" Nonetheless, Smith smiled and Willis smiled, and they shook hands. Three days later, Willis had examined the mine and wired his report: "Re Smith mine, yes."

Willis's adventures in Wilkeson continued, with various sorts of challenges along the way. When an early frost killed the grass that had fed his herd of twenty beef cattle, a key component of the camp's food supply through the winter was hungry and losing weight. "I was no dietitian," Willis wrote, "indeed I had never heard of one, but I sensed that fresh meat was an important item." Unable to pack in enough hay, Willis found a recipe for corned beef, and "we corned down our twenty head in troughs hewn from fir logs. And it was good corned beef."

Although he wrote little about it in his later autobiographical writings, in 1882 Willis found time to marry his cousin Altona Grinnell. His fieldwork continued apace. By the end of the summer of 1882, Willis had explored the region west of Mount Rainier, now part of the national park, and, "knew where good coal was to be found." Willis and the company then set out to acquire the desired land, following the letter of the law, by dispatching "stooges," usually office clerks, to the individual one-quarter sections of land (160 acres) that an individual could claim at the time with squatters' rights. In four years, on Willis's recommendations, the company acquired 10,000 acres of coal land at a cost of $90,000. The subterfuge did not offend Willis's sensibilities. "The evasion was necessary," he explained. "Otherwise the coal field could never have been explored or developed, since no one could afford the expense for the prospective value of 160 acres. It was a stupid law, enacted by a majority of Congress in the heat of the battle against monopolies."

Fall of 1883 brought the beginning of the end of this chapter of Willis's career, as the Northern Pacific Railway Company began to crumble. For a time, although the company as a whole went bankrupt, Pumpelly was able to intervene, securing a continuance of Willis's contract with the parent company through July 1884. At the time, his team was prospecting promising lands west of Mount Rainier. Willis's plans to close out his work in orderly fashion were interrupted by a telegram that reached him on April 28, 1884—"No bills paid after April 30"—leaving Willis two days to close out three and a half years of work.

Although Willis had first joined the US Geological Survey in 1879, his work with Pumpelly had taken him away from that organization. By the time circumstances pulled the rug out from his employment in spring of

1884, Willis found himself with no job and a wife to support. He had no savings or immediate prospects of further employment, so he wrote to the Survey to ask for another appointment. The month of June dragged slowly as he waited for a reply. Finally a telegram arrived: "What is the least salary you will accept?" Nudged by Altona to not "sell [himself] cheap," Willis sent back an answer of $2,400, considerably higher than the figure that had first come to mind. This time the reply was swift: "Report Washington July first."

The year 1884 cannot have been easy for Willis. Within the span of six months he not only saw three years of work collapse overnight and moved across the country with his young wife, but also saw the birth and death of his first child, Marion, in September of that year. His second daughter, Hope, would be born in 1886. This daughter survived.

By the time Willis reentered the Survey in 1884, Major John Wesley Powell had succeeded Clarence King as director. Willis's reentry into the USGS was not entirely smooth. Although his initial appointment had been signed by King in 1879, five years later Willis was "a stranger, and not cordially welcomed by an inner clique that presumed to administer discipline under the Chief Clerk." Willis did not elaborate on the reception he received, but his remark that "they did not succeed in forcing my dismissal" hints at the depths of the animosity he encountered. The tenor of agency leadership had changed with the transition from King to Powell. Willis described the former as "a man of literary tastes, a nature lover, a restless spirit, to whom responsibility was irksome," in contrast to the latter, a "one-armed veteran of the [Civil] War and daring explorer," but also "a born fighter [who] truly relished fighting his appropriation bills through Congress in spite of determined opposition." Powell's reputation as a leader had grown over the years following the war, in no small part for leading the first scientific exploration of the Colorado River.

After Willis reentered the Survey, his talents were put to use mapping coal reserves in Appalachia. Recognition of Willis's prowess as a geologist began during these years, when he saw through the literal trees to appreciate the figurative forest, developing a new understanding of the geological history of the region in spite of obstacles including the thick forests. By 1893, Willis was put in charge of the USGS Appalachian Division. With support from Powell, Willis pressed further, devising laboratory experiments using colored wax to show how folding and faulting throughout the Appalachians had developed as a consequence of broad regional compression. Willis's first book, *The Mechanics of Appalachian Structure*, published in 1894, garnered widespread interest in Europe, including in Switzerland, where geologists were working to understand similar terrains. A young professor at Harvard who would go on to be called the "father of American geography," William Morris Davis, wrote to Willis, "Your work is simply COLOSSAL. What a step it marks, what a stride."

In Willis's words, Powell "discovered that I had a knack for drafting and color combination" and in 1893 tapped the young geologist to be the editor of geologic maps. The preparation of a geologic atlas of the United States also "required lucid accounts of the facts it presented," and, as editor, it fell to Willis to review his associates' work. This, too, led to frictions, as colleagues sometimes resented his suggestions. The man who had made his way among Native Americans and cantankerous businessmen in the Wild West, earning perhaps amused respect—but respect nonetheless—from the rough-edged men he led, faced new sorts of problems in a bureaucratic environment. As Willis would soon discover, the challenge of facing starvation when early frost sets in and the challenge of finding one's way in bureaucratic waters are two different things.

Among the associates with whom Willis struggled during these years was a young geologist who was only a year younger than him, but whose early life could scarcely have been more different from his own. Although their interactions at the Survey were relatively limited, they established the character of the relationship that later defined the great quake debate. In short, and to put it mildly, these men, talented and accomplished geologists both, came to dislike one another personally almost as much as they respected one another professionally. To understand the nature of their relationship, and how their trajectories came to intersect at the USGS in the first place, it is time to meet the putative villain in the great quake debate: Robert Thomas Hill.

CHAPTER 2

ROBERT T. HILL

Every happy family is alike; every unhappy family
is unhappy in its own way.

—*Leo Tolstoy*

At his birth in 1858, in the city of Nashville, Tennessee, Robert Thomas Hill might have appeared to have had much in common with Bailey Willis, who had been born just fifteen months earlier. Like Willis, Hill was born into a wealthy, even an aristocratic, family. Like Willis, Hill's aptitudes and interests in science, geology in particular, would become evident early in life. One might have predicted parallel life trajectories for these two individuals, and one would not have been entirely wrong.

In adulthood, both Hill and Willis grew into men of less-than-towering stature. At five feet four inches, Hill was even more diminutive than Willis. Photographs paint different portraits of the two men, their generally similar dimensions notwithstanding. As a young and even as an older man, Willis is revealed in photos as a leprechaun: slightly built and energetic, a twinkle in his eye, a flashy moustache not quite concealing

Geologists at Harper's Ferry, West Virginia, 1897, including Robert T. Hill (left of center, middle; Hill, holding an umbrella, is to the immediate right of a taller man wearing a light jacket) and Bailey Willis (toward left side, middle; Willis, wearing a light cap, has his hand on the shoulder of a man seated below him, who is holding a crossed pick and hammer). (USGS photograph)

the hint of a smile. Photographs of Hill suggest not a leprechaun but perhaps more a hobbit: also slightly built and energetic, also usually sporting a (less flamboyant) moustache. But compared to Willis, Hill was a little less spry, a little more square, with more flint than sparkle in his eye.

Geographical as well as biological happenstance further shaped the personalities of these two men and dictated the course of their lives, in very different ways. The timing of their births proved critical, for one boy more than the other. When the Civil War broke out in the boys' early childhoods, one boy lived on a country estate in a part of the country that the war left physically unscathed, while the other young boy lived in a city, Nashville, that would soon feel the full brunt force of battle. In Hill's own words near the end of his life, reflecting on his childhood, "These battles of the Civil War in Tennessee did not consist of long-distance fighting directed in telephone and radio between Generals stationed long distances in the rear, but were man-to-man fights. . . . Victories and defeat," he added, "were accompanied by casualties of enormous proportions."

Robert spent his earliest childhood years in one of Nashville's elegant antebellum mansions. When he was a toddler, the family house on South Cherry Street was filled with children and games, parties and music. His mother, Catherine Tannehill Stout Hill, was later described by one of her other sons as having been "ever bright and cheerful," a mother who "led us children in our childish games, helped us with our lessons, taught us to dance, and in other ways gave joy to our lives." His father, Robert Thomas Sr., hailed from a family that had been prominent in colonial affairs. His fortunes fueled by inherited wealth if not actual business acumen, Robert Sr. became a successful merchant. Both Robert Sr. and Catherine were musical: he played the violin, she the piano. But unlike his wife, Robert Sr.'s temperament was less than even: his son Jesse later described his father as prone to "violent fits of temper." The marriage, nonetheless, was fruitful. The couple had eight children, the eldest born in 1844 and the last child, Catherine, in 1860; Robert Hill Jr. was the seventh child. Unlike the youngest child in the Willis household, Catherine survived, and for a brief time she became Robert's closest playmate. One of the middle children, Josiah (Joe), born in 1852, was too much older to be Robert's close companion in early childhood; Joe was, however, the family member who would one day most profoundly shape his little brother's life trajectory. The large household included Robert Jr.'s maternal grandparents and a number of slaves. Whereas Bailey Willis was raised by abolitionists, young Robert was nurtured and literally nourished by slaves, breastfed not by his mother but by a wet nurse, a slave named Mary.

Through Robert Jr.'s early years, the surficial trappings of his life were largely similar in many ways to those of Bailey Willis: secure, sheltered, and materially comfortable. But whereas, by all accounts, Bailey Willis's mother was emotionally as well as financially sturdy, Hill's parents' psyches increasingly showed signs of cracks around the edges. What happened during Robert Jr.'s earliest few years did not take definite root in his mind. Later in life, Robert retained no clear memories of the first few years of his life, no recollections of gay music or happy times. In later years he wrote of just a few faint memories from the first three years of his life: a recollection of the fine fruit trees in the family yard, the fragrance of lilac bushes, the promise of a watermelon set to cool in a tub of cold water. The first tune he specifically recalled having heard was Pleyel's funeral march, along with the "muffled drums and solemn visaged soldiers" he observed as a child of four. By this time, the war had begun to take a heavy toll on Nashville, having just claimed the life of a beloved native son who had risen to the rank of general in the Confederate Army.

Although the Hills owned slaves and Robert Sr. and Catherine reportedly "did not believe that ownership of slaves indicated partnership with the devil," both were among the minority in the South who had some sympathy for the Union, reportedly supporting the cause of abolition by legal methods. "There is every reason," Robert Jr.'s brother Jesse would later write, "to believe that the actual breaking out of war broke all their hearts." The eldest children stood squarely on the side of the Confederacy and rushed to join the cause. None of the Hill children met the official minimum enlistment age of eighteen, but official rules fell by the wayside in the face of wartime exigency. Many younger boys, with commitment to the Confederate cause perhaps buoyed by dreams of coming home as heroes or simply a yearning for a more interesting life, passed themselves off as older, and many younger boys joined legitimately, serving in noncombat positions. Robert's eldest brother, Thomas, enlisted in the Confederate Army at the age of fifteen; at just thirteen, the next-eldest Hill boy, Samuel, became a drummer boy. Even the next-youngest son, Jesse, ran away from home twice to join the army, but was not successful. Catherine's father fought for the Confederacy as well. Robert Sr.'s role in the war is unclear: by Jesse's later account, he and his wife had not been supporters of the Confederacy.

Initially an important administrative center for the Confederacy, Nashville quickly became a literal battleground; by 1862, the city had become a Union stronghold, ruled by martial law. The Union confiscated many of the city's homes for use as army headquarters, among them the Hill home on South Cherry Street. The Hill family, including the remaining six children, moved to a small rented house. Robert Sr.'s business collapsed, and with it the family's fortunes: on one occasion dinner was just a bowl of sweet potatoes. Genetically Robert Jr. was probably not destined for great height; although his father was tall, Jesse later described their mother, along with most of her family, as having been much below average height. Whatever adult physical stature Robert's genes might have had in store for him, the cause cannot have been helped by early years of malnourishment.

The war continued to take its toll on the Hill family. Thomas and Samuel survived the battles of Mill Springs and Fort Donelson, but had been taken prisoner, although the family would not learn of their fate for many months. They learned that Sam had been sent as a prisoner to Camp Douglas in Chicago. "Our anxiety," Hill's brother Jesse later wrote, "was almost as great as lack of knowledge of [Sam's] fate, because we then knew what fate his captors might have in store for him, for at that time we had begun to regard the 'Yankees' as monsters." Grim circumstance increasingly affected Catherine Hill, who began to slip into depression. In mid-1863, by Hill's later account Robert Sr. developed terminal pulmonary disease, brought home from a battlefield on the Cumberland, following a stint of employment with the US Quartermaster. He spent the rest of his life bedridden. It fell to thirteen-year-old Jesse to become the family breadwinner; the US Quartermaster Department offered him a job as a messenger, paying twenty-five dollars a month, which he accepted gratefully. In September 1863, Catherine's mother died, and Catherine suffered what was then commonly referred to as a nervous breakdown. Later in life, her youngest son would retain only a single memory of both of his parents. The recollection of his father was more-or-less benign, Robert Sr. telling his son to "wait for the wagon" after he had impatiently clattered his spoon at the dinner table. Hill's sole memory of his mother was less benign: an image seared on a young boy's consciousness of his once-pretty and gay mother screaming for her children, left behind as a Yankee military squad dragged her away. Hill's father, realizing that "his own demise was inevitable," had arranged for her confinement. Robert's father indeed died not long thereafter.

Although some brief biographies say that Robert Jr. was orphaned at age five, and Robert himself sometimes described himself as such, his mother lived for many more years. But all of her remaining thirty-seven years were spent in the Tennessee Hospital for the Insane, where she had been admitted in November 1863; her children never saw or heard from her again. Many years later, Hill learned that she had been diagnosed with "acute delusional insanity," a chronic case with "gradual weakening of all mental faculties." Robert and his siblings might not have been orphans in the strict sense of the word, but they might as well have been. At the age of five, the world around him still embroiled in war, the young boy stood outside with sister Catherine, feeling very alone. Robert soon bade a tearful good-bye to his baby sister, sent with one of their older sisters to live with an uncle in a different county. Along with Jesse and older sister Sallie, Robert was taken in by their paternal grandmother, Sally Wood Hill.

The next chapter of Robert's life began in another grand mansion—one that had been spared from confiscation during the war—but with a different tenor of life than the one he had been born into. The children received instruction ranging from proper manners to spelling, but their days were shaped first and foremost by religious instruction and devotions. "The days at home opened and closed with hymns, Bible readings and long prayers. Everyone was awakened by the ringing of a huge hand bell, and woe to him who was not promptly at hand at prayers." Sunday devotions began with Bible readings at home, followed by Sunday school, followed by church. Hill later described "long and dreary service[s] with hymns like, 'There is a Fountain Full of Blood'" and sermons that included "much Bible pounding and much denunciation of dancing, card games, fiddle playing, theatre going, and other pleasures of life." After lunch, the family returned to church for a "singing," which the children actually enjoyed: they were allowed to smile and sometimes to walk home with their friends.

At the age of five, Robert found himself with a different sort of primary caregiver than Bailey Willis grew up with: devoutly pious, severe and strict, with deep, abiding enmity toward the Union and the military governor, Andrew Johnson. Grandmother Hill ruled the roost, but not, by Robert's later recollection, warmly. Robert later described life in Sally Hill's household as a "volley of repressive don'ts." The extended household was not entirely devoid of warmth for a young boy. "The nearest substitute I ever had for a mother," he later wrote, was one of his father's sisters, Aunt Emmeline, also part of the extended family in the household. The people who had made up the fabric of his early childhood years had vanished altogether. Hill later wrote that a former slave known as Aunt Jane, "a bright spot in the recollection of an orphan of the Civil War," was the only one who later sought him out after the war, referring to him as her baby. Jesse, old enough even as a young teen to be considered a man in a society where so many men had not come home, eventually fled the household to make his own way in the world. His precise trajectory is unclear; for a time he remained employed by the US Quartermaster's office and continued to send money to his grandmother to help support his younger siblings. Robert and Sallie were fed and clothed; the roof over their heads was sound, and their educations resumed when local schools reopened. But it proved to be a dreary time for the children as well as the city as a whole. Beyond the Hill household, surviving men returned to Nashville defeated, without money, their opportunities scarce. At school, the children were taught by teachers brought in from the northeast, who brought with them "the prevailing northern idea that the art of speech and every domestic custom down South was wrong" and no hesitation in using corporal punishment to make their young charges speak what the teachers called proper "Yankee talk."

Around the age of nine, Robert felt obliged to contribute to the meager family income by taking on a paper route that required him to wake in the wee hours of the morning to complete his rounds before going to school. After five hours of work, he arrived at school already exhausted. The severe, berating nature of the instruction did little to kindle his spirits or interest in the subject matters. His attentions turned to mischief of various sorts: playing hooky from Sunday school to venture into the back door of a beer saloon ("just for the wickedness of it"), and generally taking an "almost fiendish delight in breaking every mandate." He secretly learned to play poker "just because I saw my sporty young elders doing it," perhaps planting early seeds for a lifelong penchant for gambling. His intellectual energies did find one outlet: the armfuls of books he pulled from the family bookshelves in the evening, including classic novels and Shakespeare plays. He did not find books by the likes of Darwin and Huxley; such books had been hidden from the children for fear of "polluting [their] minds with disbelief." But the books he found, he devoured. At some point during the postwar years, young Robert also found an old iron coupling pin lost from an artillery wagon, which at one point he used, in standard little-boy fashion, to pound a rounded rock into fragments He was surprised to find a symmetrical structure inside the rock; he had no way to know at the time that the round rock was in fact a fossilized coral head.

By the time he reached his teens, by Hill's later account, there was little love lost between him and his extended family. His schooling had ended after sixth grade, but a professional niche had failed to materialize. Robert was viewed as incorrigible for not altogether unfair reasons, and the family despaired of his career prospects. Perhaps they loved him and wanted only the best for him; one imagines that the devout, dutiful family did their best in the face of the pressures of the time. Following Sally Hill's death in 1893, a lengthy obituary in the *Christian Advocate* held her up as a paragon of Christian virtue, a woman who "searched the Scriptures diligently, and as steadily did her hands hold the distaff." Of her piety, the obituary had much to say: "Her children, and their children, were taught to respect the men whom the Church had raised." The article also painted her in warmer tones than her grandson later would: "Candid, sympathetic, and approachable, she drew all hearts unto her, and shed over all the benign blessings of her sweet spirit." For his part, Hill did not feel like the recipient of benign blessings; as a dreary progression of days dragged by, he felt like little more than an extra and unwelcome mouth to feed. His older brother Jesse later offered up a different, more charitable perspective. In 1932 he told his brother that he had always had a feeling of gratitude toward their grandmother's family for taking them in when they most needed it. He moreover had "never believed it was a financial burden" on the family, since his own rations and entire salary "more than compensated for the board of all three of us." There might be reasons other than temperament why Jesse's perspective differed from that of his younger brother: Jesse was not part of the household during his younger brother's later years there, the years that Robert himself would remember most clearly.

In the end, Hill's older brother Joe, who had been sent to live with an uncle after the children were parceled out among relatives, provided Robert's opportunity to escape from the Hill household. Joe had made his way to Comanche County, Texas, some years earlier, eventually working as the editor and manager of the town newspaper, the *Comanche Chief.* In 1874, Joe sent a letter to his brother, inviting sixteen-year-old Robert to join him and work as a printer. His family begged him not to go—not, he felt, because they would be sad to see him go, but because working in such a "lowly profession" would bring disgrace to the family. Among the still-proud Southern aristocracy, only certain professions, not including the newspaper business or other "trades," were considered respectable. Determined to grab the lifeline he'd been offered, he argued that any whiff of disgrace from Texas would not reach Tennessee. The field of study to which he would later devote his life was, of course, nothing short of blasphemy to deeply pious family members. In late 1874, his fare paid with money gifted by a family friend, he boarded a train for Texas and the life that awaited him on the frontier.

A lumbering steam engine pulled the train as far as East Waco, where the rails ended. The fresh-faced, barely five-foot four-inch Tennessee boy found himself in the company of rough-hewn men. The horses, curiously, were not Texas-sized but, rather, mustangs, smaller and scrappier than the big thoroughbreds Hill had known in Nashville. He also found himself more than a hundred miles from Comanche, with a single dollar in his pocket. After a cold and hungry night on the streets, he found transportation on an ox-drawn wagon full of timber but soon calculated that he could walk faster than the plodding beasts and set out on his own, not realizing that people in this part of the country viewed a person on foot as a horse thief in the making.

Hill's childhood might have been different in many ways than Willis's; he had certainly known more hardships, including a firsthand look at the horrors of war. But Hill had also grown up in fundamentally sheltered waters; while his extended family might have felt less than ideal to Hill, their household was the antithesis of rough frontier life. Hill later described himself in early childhood as a Little Lord Fauntleroy kind of boy. As had been the case for Willis, Hill's horizons expanded quickly as he took steps into the larger world. Hill soon rejoined the wagon, walking beside it and cleaning mud from the wheel spokes. Rain fell steadily during the miserable two-week journey, leaving the party with soaked clothing and bedding. At last, on Christmas Eve, they reached Comanche. For the first time since he left Tennessee, and for the last time in years, Robert slept in a proper bed that night. For the next seven years he slept on a pallet spread on the floor of a print shop, after working ten-hour days as a "printer's devil"—seven years during which he "seldom saw butter, ice or a bathtub," or long prayers or dreary church services (although he did go to a local church). He had put about as much distance between himself and his Little Lord Fauntleroy childhood as a young man could have managed without actually leaving the planet.

The next, and very different, chapter of Robert's life began with him working as a newspaperman in a frontier town. The job description of a printer's devil basically meant doing anything that needed to be done in the *Comanche Chief* office, from cleaning to setting type to writing. Back in Nashville, Hill had been accustomed to a life with little youthful companionship. In the hardscrabble frontier town he was even more isolated. He read newspapers to stay abreast of news in the outside world; with limited local amusements, he continued his boyhood habit of poking at rocks. In arid central Texas he found a great deal more rocks to poke than he had back in Tennessee, and poking at them turned up more interesting things, including different types of fossils. Intrigued by his findings, Hill asked the town druggist, who ordered and sold books, to order a geology textbook for him. That book, by James Dana, was at the time a classic text; years later, Hill himself would help revise a later edition of the book.

During his Comanche years, Hill also took the opportunities that presented themselves to expand his literal as well as intellectual horizons. He sometimes joined businessmen from Dallas and Fort Worth, known locally as "drummers," who craved companionship on their long rides between settlements. In the summer of 1876, young Robert joined a team of surveyors heading to west-central Texas, lands of reddish-brown rock with occasional layers of white gypsum, now referred to as redbed country. Among the thirty-odd members of the team was a former colonel, an educated and literary individual who found common interests with the young man. Talking into the night, the colonel taught Robert the constellations, bright in the frontier sky. During those nights, Robert later wrote, he felt "the first inspiration to study and contemplate the things of nature." On the plains of west-central Texas, the Sunday school boy from Tennessee felt himself evolve, becoming "a continuous seeker after knowledge."

Hill's burgeoning curiosity about the natural world soon combined, perhaps inevitably, with the grandeur of the setting in which he found himself. Later in life Hill would describe it as the "greatest loss of [his] life," the eight years of his youth that "should have been spent in college prep." But even Hill acknowledged that his years on the Texas frontier had provided their own sort of education, a "golden opportunity to acquire a 1st hand acquaintance of pristine nature, and to ascertain the bedrock valuation of men & women." The rugged land, studded with flat-topped mesas, was so different from the green hills of Tennessee. He discovered and climbed nearby Round Mountain, where rock strata and fossils were on especially clear display, and which provided a vantage point to survey the otherwise mostly flat terrain. What, Hill wondered increasingly, had shaped the land in such a way, leaving the tops of mesas so flat? How did one account for the shells? From his one geology book, he learned that the shells he'd found in a layer of rock were marine fossils. But "how on earth did they get raised a thousand feet above the sea?" The book provided no answers, and an acquaintance from the local church he attended did not respond well to his suggestion that perhaps the fossils were the remains of animals that had lived in an ancient ocean. As Hill pondered such questions, he began to collect rocks, going so far as to advertise in the newspaper, inviting the citizens of Comanche to bring him "any queer or unusual kind of rock." Amused and bemused, some residents began calling him the "rock boy" of Comanche; to others, he was the "crazy printer."

In 1877 Hill left Comanche, this time as a cowboy herding cattle and "learn[ing] the language of the trail." Hill and the cattle arrived safely in Dodge City, Kansas, but there met with "bad fortune." While he did not elaborate on the nature of his bad fortune, one can hazard a guess based on a reference he made in his later years to his "gambling spirit." In any case, Hill found himself miles from home and dead broke. Seeing a cattle train about to pull out for St. Louis, Missouri, Hill offered to trade cattle-tending services for passage to St. Louis, a deal that left him still broke but able to visit an aunt and uncle in that city. When the aunt asked him to turn his money over to her for safekeeping, he told her that he had left his wallet at the wagon yard; she told him to get it the next time he was there. In due time he went to the yard, but rather than face his aunt again, he hopped on a freight train heading south. The ride did not last long; he was discovered early the next morning and booted off the boxcar, landing near a farm where he found work helping a farmer bag corn. He left the farmer and his corn behind, taking an even more backbreaking job: hauling cross ties. He soon left this job behind as well, catching another freight train, planning to get off at the Texas border. Instead he fell asleep. He awoke to find himself in the train yard at Waco, "stiff, dirty, hungry, still broke, but about 100 miles from the printing office at Comanche." Eventually he made his way back to Comanche and, temporarily, the printing office.

Hill's interest in geology kindled, in 1880 he made an initial attempt to seek further education. He set out for Nashville, "where Vanderbilt University had been started and where I had influential relatives." He traveled through Memphis, finding work for a time with the local newspaper. He eventually arrived in Nashville to an initially warm welcome, but "when I revealed my desire to earning [*sic*] my way through college by manual labor as a printer . . . the welcome became frigid." The little boy who had left there eight years before in his Lord Fauntleroy collar was not welcome, he felt, when he "came back in jeans trousers, hickory shirt, and brogan shoes." Dispirited, Hill once again made his way back to Comanche.

Back on the frontier, Hill found a new companion in the Comanche town barber, Emil Ulrich Wiesendanger, economically known as Wies. A Swiss immigrant, Wies was an anomaly in the Old West, a man who, like young Robert, found refuge in books. By Hill's later account, Wies was the one who first set him on a trajectory that led to his college education. Eventually there would be two different versions of the story. According to one version, in late 1881 an animated discussion about the possibility of a fourth dimension ended with the conclusion that Weis "did not know a damned thing about it" . . . and Hill didn't either. Weis asked Robert, "Why don't you go off to college somewhere where you can find out about it?" In the other version of the story, young Robert asked Wies how he might learn more about rocks, and Wies replied that he could go to college. A scrap of paper in Hill's notes suggests that the latter story was the correct one, although the former was more frequently repeated. In either case, by then he was twenty-three years old, his formal education had ended with sixth grade, and he was living far from the nearest institution of higher learning; college might have struck Robert as about as attainable to flying to the moon . . . and yet, perhaps not impossible. A Vanderbilt education might not have been in the cards for him, but reaching out to contacts at eastern newspapers, Robert learned that Cornell University had recently been founded on the principle that "any person can find instruction in any study" and that it was possible to work one's way through school.

The events set into motion one way or another by a Swiss barber on the Texas frontier culminated with Hill's boarding a stagecoach in February 1882, wearing cowboy boots and carrying a suitcase full of rocks, to embrace his future. A month later, the journey brought him across the Mason-Dixon Line into what he had been raised to view as nothing short of enemy territory. It was in many respects a different world. Having left behind the green grass and wintertime flowers in Texas, Hill arrived in Ithaca, New York, in the middle of a snowstorm. He had arrived: opportunity beckoned, his chance at higher education, in some ways his first opportunity for any first-rate institutional education. But if ever an aspiring college student was ill prepared for the road ahead, Hill was that student. Of immediate concern was the fact that he had no coat or other winter clothes. Having borrowed money to pay for the trip, he had no money beyond ten dollars cash in his pocket and little family support behind him. He moreover arrived with so many deficiencies in his educational background, his acceptance had some strings attached: he would need to demonstrate mastery of Greek and Latin, among other subjects, before he could graduate.

At Cornell, Hill also ran into sensibilities that sometimes seemed more like those he had fled in Nashville than those he had developed during his seven years in Texas. At a required lecture on physical hygiene, the class of boys, almost all younger than Hill's twenty-three years and far more privileged, was informed that committing certain untoward acts— gambling, drinking, and so forth—would be cause for expulsion. The measure of irony invoked an intense reaction in Hill, who by then had experienced far more of life's rough edges than the speaker. "What t'ell business have we redblooded folks," he wanted to ask, in this "mamby-pamby, plus-fours, tenderfoot co-ed place." By this time, Hill's years in Texas had, it seemed, thoroughly overprinted his own Sunday school days in Nashville. Hill realized that he was once again an outsider, that he would have to shed the toughened skin he had developed during his years on the frontier and "be a little child again." Cornell would, however, prove to be educational for Hill in many ways. During his student years, he found himself in a university drinking establishment with some of his classmates, seated next to none other than Mark Twain. Hill observed to his companion that these students, with their "tailor-made clothes, white collars and fraternity pins . . . [were] . . . about as bad as the old-time roughnecks" he had encountered in Dodge City. Twain retorted dryly, "A damn sight worse."

Against all odds, Hill not only survived but thrived in his new environment. Once again, life was not easy. For a time he bunked with two other boys in a tiny building that he described as an outhouse and took whatever jobs he could find to support himself and his education. When that wasn't enough, he accepted some assistance from his brother Jesse and took out student loans. Still, it was nothing short of a revelation for a young man whose intellectual talents and interests had never been nurtured, to be in a place where people devoted their lives to the pursuit of knowledge, unfettered by strict interpretation of and devotion to Scriptures. He excelled at his studies and soon began to take on leadership roles. Later in Hill's life even his closest friends acknowledged that he had a Texas-sized contrary streak, but he was never antisocial. Throughout his life he forged deep friendships and earned the respect of many who admired his intellect and appreciated his company. At Cornell he also survived his first experience with integration: "A Negro student sat in the same class room with me and I did not drop dead." A further revelation came in his first encounters with intelligent, assertive, independent young women. These women struck him as a resounding departure from the two types of women—dutiful homemakers and bar girls—he had known on the Texas frontier. Among the women he encountered was his future wife: Jennie Justina Robinson, attending Cornell for postgraduate studies in history.

Cornell had much to teach Robert Hill, but even there, nobody could teach him about the subject that had captured his interest during his days on the frontier: Texas geology. Encouraged by a geology professor, Hill undertook an independent course of study, relying on published literature to understand the state of knowledge at that time. The information he found was fragmentary and scant. For his undergraduate thesis, he wrote a first-ever compilation and synthesis of geological information about Texas. By his junior year, word of Hill's interests and talents traveled south to Washington, DC, reaching John Wesley Powell. The year was 1885, one year after Bailey Willis had returned to the US Geological Survey. At the helm of the still-young agency, Powell needed more energetic, talented young men to carry out his agency's mission. Powell offered Hill a position in June 1895; Hill accepted, planning to take a leave of absence from his education, but still hoping to graduate on schedule with his class in 1886.

For a time, Hill remained confident that his new position would not derail his formal education. In the end, the university stuck to its guns regarding requirements in Latin and Greek that Hill had not met before he started work with the Survey and did not have time to complete once he had. Hill pled his case with a note of petulance: "I failed to prepare myself," he wrote, "only because the University failed to allow me to do so." As his graduation date approached, Hill held out hope that he would be able to graduate with his class. In the end, the matter was settled by a letter that Hill received just an hour before the graduation ceremony: he would not be allowed to graduate that day.

Hill left Cornell a final time, feeling humiliated, without the degree for which he had overcome so many obstacles to pursue. Whether it had been the university's fault or his own might not be clear, but in any case he parted company with the university on a bitter note. Rapprochement with his alma mater would come later. But even with a taste of bitterness in his mouth, he might have appreciated that he was leaving with something worth far more than a piece of paper: the job, it seemed, of his dreams.

CHAPTER 3

INTERSECTING ORBITS

> Perhaps a man's character was like a tree, and
> his reputation like its shadow; the shadow is what
> we think of it; the tree is the real thing.
> —*Abraham Lincoln*

When Bailey Willis returned to the US Geological Survey in 1884 at the age of twenty-seven, he had landed in a position for which he had been groomed his entire life not only by his innate talents but also by life circumstances. When Robert Hill began his career at the same agency in 1886, at the age of twenty-eight, he landed in a place that he had fought and scratched his entire life to be. Born just a year apart, with similar talents and energy but on opposite sides of the most important political divide in American history, the two men's early trajectories could scarcely have been more different. Yet by 1885, Hill and Willis had both ended up in the same place, eager to make their mark in the same scientific field. As young men with keen interest in sorting out the geological landscape, they could not have arrived in their field, or government agency, at a better time. There was an entire country to be mapped, with a new government agency created to take on the challenge. The big-picture scientific questions had to be sorted out: What were the major physiographic features? How old were they—including the flat-topped mesas that Robert Hill had wondered about on the Texas frontier? And how did they fit together?

Of primary concern to the country, and therefore the agency, was the mapping of mineral and fossil fuel resources. Where did the coal reserves lie? The ores? The oil and gas? In Hill's later words, the agency found political support from those keen to "develop [the South's] economic resources." Although Hill's and Willis's swords were destined to cross throughout their lives, Hill later gave a nod to the latter man's scientific contributions, noting that "Major Powell repaid [those who had supported the USGS] with the splendid results of Willis and Hays in the study of the coal fields and iron ranges of the Appalachian region." Geological acumen was especially critical to guide the search for fossil fuels. About 70 percent of the world's oil, for example, is found in strata of Mesozoic age, the "age of dinosaurs" that spanned the period roughly 252 million to 66 million years before the present. Over the ages, oil collected in reservoirs where the subsurface geology creates structures known as traps. The search for fossil fuels therefore requires, essentially, mapping of subsurface geology. Today scientists can investigate subsurface strata by recording waves created by explosions to create an image akin to a CAT scan of the earth's upper few miles. In Hill's and Willis's day, it fell to geologists to infer three-dimensional structure from two-dimensional observations—the features that can be mapped at the surface. More important than oil in the late nineteenth century were the country's rich coal reserves. It would fall to the nation's entrepreneurs to exploit the natural resources and build great empires. The basic science fell to a different breed of men, including Bailey Willis and Robert Hill, men who were naturalists and explorers at heart.

Today we understand three-dimensional landscapes within the paradigm of plate tectonics, which explains how landscapes have been shaped over geological time. We know that heat generated by radioactive decay in the earth's core causes slow convection of the overlying mantel, which drives the motion of overlying rocky plates that make up the planet's thin outer crust. This basic scientific framework would not be worked out until the mid-twentieth century; by the end of the nineteenth century, geologists faced the more basic challenge of characterizing and understanding the rocks themselves. In effect, it fell to Hill's and Willis's generation to map the puzzle pieces that plate tectonics would later put together. It was a grand challenge after a bright, keen, energetic young scientist's heart.

In retrospect at least, it would have been a heady time to be a young geologist working for the federal agency tasked with, literally, surveying the geology. Ironically, however, neither Hill nor Willis met with altogether smooth or happy sailing during their early years with the agency. Willis chafed within a bureaucracy in which, in his view, a new cadre of Powell men did not take kindly to him as an employee brought to the agency initially by Clarence King. Hill, by contrast, was a Powell man, plucked from his studies at Cornell by the major himself, who became nothing short of a personal idol to Hill. Yet almost immediately Hill, too, ran into difficulties. Whereas Clarence King had dispatched Willis to work with Raphael Pumpelly, Hill drew a different straw with his first supervisor. Appointed as an assistant paleontologist, Hill was dispatched by Powell to work at the National Museum of the Smithsonian Institution, under the direction of fifty-nine-year-old geologist Dr. Charles Abiathar White. By that time, scientists knew that fossils, which had fascinated Hill as a young boy, provided a Rosetta stone with which geologists could establish the ages of different rock units. When Hill began working at the

USGS, his overwhelmingly positive associations with Northerners at Cornell had dispelled, or at least pushed to a back burner, the deep biases about Northerners that had been instilled in him during childhood. In White, who he referred to as the "Ungracious One," Hill found a man he considered to be a living, breathing embodiment of every worst stereotype he had heard, a "genuine specimen of my uncle's 'blue-bellied' ideal." Hill later wrote, "He showed me a disposition as gentle as that of a snapping turtle, and banished me to the company of the Fiji spears, shields, and malodorous smells of insect and snake preservatives . . . to sort out collections of fossil shells." Within the confines of these quarters, in October 1885 White granted Hill permission to spend not more than one-half day each week for general study of his own choosing.

For his part, White was a successful, highly prolific geologist, described by his contemporaries as alert and engaged until nearly the end of his eighty-four-year life. Perhaps there were reasons other than prejudice behind his initial reception of Hill; perhaps Hill took offense where none was intended. But at least in Hill's view, prejudice against him as a Southerner had fueled White's outrage at the seventy-five-dollar weekly salary that Powell had offered Hill. According to one of Hill's colleagues, White had demanded that twenty-five dollars a week be cut from that amount and added to his own salary and proceeded to relegate the talented young geologist to dusty museum chores. The experience, and others that would follow, rekindled Hill's feelings about Yankees and their view of those from the South.

Embittered at the start, Hill considered leaving the Survey, but decided to stay. By July of the following year he received a more congenial assignment, liberated from the malodorous corridors of the Smithsonian to undertake a three-month field study in his adopted home state of Texas, with instructions to submit monthly reports. The trip gave him an opportunity to investigate further the nature of geological strata through-out the state and understand how they fit together with strata elsewhere in the midcontinent. He came to see that prevailing views on the subject had been based on misperceptions. In short, he confirmed his earlier belief that parts of the Texas landscape were much older than generally thought, upward of 100 million years old. It might not sound like a block-buster result that the lower Cretaceous was discovered in Texas, but in geological circles at the time, it set conventional wisdom on its ear. Hill also presented what nobody else had done, a geological column showing the true stratigraphic relations between the formations in Texas. In short, he was the first geologist to identify the state's major rock units, past and present, and describe how they fit together. Hill's work, while immensely gratifying to him as a geologist, led to further run-ins with White, who saw it as his prerogative to summarize and present the results that Hill had done under his broad supervision. White prepared his own presenta-tions, drawing from the detailed reports that he required Hill to submit regularly. From the start, however, White struggled to read these reports; he wrote several letters to Hill in the field, requesting that he "give much more attention to the cleanliness of [his] handwriting." The many per-sonal letters that Hill wrote throughout his life reveal handwriting that, while not textbook perfect, was generally mostly legible. It remains unclear, then, whether White's complaints were a consequence of his own prim schoolmarm sensibilities or a personal bias toward Hill, or perhaps of writing that edged toward illegibility on purpose. Wherever the truth lay, from the start Hill's personal enmity for White ran deep.

The situation in which Hill found himself has played out many times within the halls of science: senior scientists staking claim to work sub-stantively led by a student or young scientist under their supervision. The extent to which this is fair on the part of the senior scientist depends in part on the degree of guidance and supervision they provide. In modern times, when students or young scientists undertake substantial work or carve out important new research directions under general supervision, it is customary that their leadership of the project be acknowledged, for example with first authorship of the initial scientific paper presenting the results. Even today, however, a riddle sometimes makes the rounds: How many scientists does it take to make a scientific discovery? Answer: Five—a postdoctoral researcher to write the proposal, three graduate students to do the work, and a senior scientist to take the credit. A more rigidly hierarchical view prevailed in the nineteenth century, but also, Hill worked for a governmental agency with a hierarchical organizational structure.

Hill began to withhold his notebooks from White, concerned about misappropriation of his "life's work." White then went to Major Powell to demand that Hill be fired. Powell wrote to Hill, "Possibly you are not aware that the custody of geological notebooks . . . is regulated not only by custom, but by official 'Regulations of the United States Geological Survey' which specify that such documents are the property of the government, and under the control of heads of the divisions until the material which they contain shall have been put in more permanent form." The rebuke brought Hill up short; he wrote to Powell to express concerns about unjust accusations and about publication of his own paper. Powell replied, "I am sorry that you completely misunderstand the state of affairs in Washington with respect to your relations with the Survey in general or Dr. White in particular. No charges whatsoever have been made against you, and there is not the slightest occasion for asking for a suspension of judgment, since there is nothing of importance to adjudicate." Powell's letter continued, however: "In so far as adjudication was required in this matter I have decided in your favor." Powell went on to offer words of reassurance to his prickly young deputy: "You have no occasion for uneasiness or anxiety so far as your relations to Dr. White are concerned, [and] I am desirous of promoting your interests in every possible way." Powell's adjudication paved the way for Hill's own paper on his discoveries to be published in a scientific journal. The paper, "The Topography and Geology of the Cross Timbers and Surrounding Regions of North Texas," was published just a month later, in April 1887, in the *American Journal of Science*. As with Willis's early publications, the geological community took note.

The same year that the paper was published, Hill also reached rapprochement with Cornell University after he found time to take a Latin exam to fulfill the requirement that had held up his graduation a year earlier. In a letter dated June 12, 1887, a Latin professor who had taken a stand against Hill's graduation earlier, wrote, "It gives me great pleasure to tell you that you succeeded in passing the examination, and to feel that nothing now stands in the way of your degree. I cannot congratulate you yet on a very extensive knowledge of Latin, but enough is in this case as good as a feast." Later that month, Cornell University sent him his diploma, formally granting him a bachelor of science degree with special honors.

Hill moved forward with a career that would ultimately lead many to regard him as the father of Texas geology. Although he worked on and in other places, he expanded on the insights of his early days to sort out the geological structure of the place that had become, for all intents and purposes, his home state. He also made occasional forays into neighboring states. In 1887 Powell set up a cooperative agreement with the Arkansas State Geological Survey, dispatching Hill to lead the USGS's part of the work. In Arkansas Hill reported to John Casper Branner, at that time head of the agency.

Branner, who will make a number of cameo appearances in this story, had more than a few things in common with Robert Hill. Like Hill,

Branner was the son of a merchant; like Hill, he had been born in Tennessee, but in 1850, eight years earlier than Hill. Like Hill, through his childhood Branner devoured the books he could find. When the Civil War broke out, Branner was closer in age to Hill's brother Jesse than to Hill himself; like Jesse, he tried twice to enlist in the Confederate Army, but, at thirteen, was turned away. As had been the case for Hill's family, the war would not be a faraway concern for the Branners: one battle was fought on their family farm, although in their case the family home was spared. After the war, Branner remained in Tennessee, finishing his education in due course and spending a year at Maryville College before he, too, headed north to attend Cornell University. He arrived at Cornell in 1870, more than a decade ahead of Hill, and left four years later; in another parallel with the younger man, Branner left to work as a geologist without completing his degree. By 1887 he had been appointed state geologist of the Arkansas Geological Survey. While Hill's correspondence with a cousin suggests that he was unhappy to have been "loaned" to Arkansas, at least in his mind leaving Charles White to reap the rewards of Hill's earlier work in Texas, Branner became Hill's lifelong confidante, a fellow Tennessee Confederate with whom Hill found a natural kinship.

In the closing days of 1887, Hill codified his personal kinship with Cornell classmate Jennie Justina, the couple moving to Arkansas after their wedding on December 28. It was initially a happy union, Jennie accompanying her husband's forays into the field and becoming an avid rock hound herself.

Although the 1880s were productive years for Hill, through this time he yearned to return to Texas, to "work among the only people whom I have ever considered as my own." In the fall of 1888, opportunity presented itself when a regent of the University of Texas invited Hill to apply for a professorship, a geology position that had, in fact, been established specifically for him. Hill accepted eagerly, notwithstanding funding limitations that prevented the university from offering him a full professorship. Powell wrote a letter of recommendation for Hill, indicating that the USGS might be able to contribute to the publication of Hill's future work. From the start, however, Powell voiced doubts. "That is no place for you," he told Hill; "you will come back, and the door will always be open for you." Powell, a scrappy character in his own right, is among a small handful of colleagues, including John Branner, who were able to appreciate Hill's talents without tripping over his less endearing traits.

Hill arrived in Austin to join a young institution, founded only six years earlier. The faculty consisted of fifteen professors, only three of whom, including Hill, were in the natural sciences. By the time Hill arrived, a total of 278 students were enrolled, 30 of whom took classes with Hill during his first year. The university might have been new but, especially in contrast to the liberal approach Hill had known at Cornell, the Texas institution had adopted an old-school approach to education. Hill, however, brought his Cornell sensibilities to the position. He did not assign required textbooks but, rather, "tried to use the objective method of first getting the student interested in an object and then have him find out all about it for himself." As a classroom instructor he could be disorganized and unconcerned with what he saw as trivialities. The guts and the heart of his teaching methods lay in geological fieldwork. Among the university faculty, he alone made fieldwork outside the classroom a centerpiece of his classes. "There is but one geological laboratory," he wrote in an article published in *Popular Science* in 1891, "and that is the great out-of-doors."

The liberal pedagogical approach he had relished as a student at Cornell went over like a proverbial lead balloon with the university administration in Texas. Hill's colleagues did not approve, and his students also struggled with Hill's haphazard approach to classroom instruction. One student, Thomas Ulvin Taylor, who in 1907 became the first dean of the University of Texas, later wrote that "Hill had never taught when he arrived at the University, and the routine regulations of class hours, examinations, and schedules were irksome to him, and he chafed under Faculty regulations." But his students did appreciate Hill's unorthodox approach. Hill's penchant for freeing students from the classroom to do laboratory work and fieldwork. "He was eager to turn loose the pack on the trail of unknown game in the ages gone by." Taylor noted with admiration that Hill was as great in his field as the best of the other faculty were in theirs. For his part, Hill embraced with passion his role as an educator, seeking not necessarily to train the next generation of experts in the field, but to instill an appreciation of geology in all students. "When we lay by our icthyosauriaus [*sic*] and useless crystals for advance study," he wrote, "and reach the ordinary and not the extraordinary features of the earth, geology will be appreciated, and every farmer, every builder of homes, every drinker of water, will learn that upon a knowledge of its simple laws his success depends." With these words, and many others he would go on to write, Hill revealed an abiding view of geology as a science that is important because it is relevant to everyday life and economic prosperity. What's more, at a time when such was scarcely the norm in field geology, Hill's egalitarian views extended across gender lines: female students were as welcome in his classes as their male brethren. In his view, geology was a science for everybody.

Inevitably, as Powell had foreseen, the devoted, passionate, and sometimes pugnacious Cornell-educated professor found himself increasingly at odds with the university. In the later words of Thomas Taylor, "Robert T. Hill was too full of energy, too full of enthusiasm for the cloistered halls of a university of olden days." The university might have made its peace with Hill's unorthodox approach to education had Hill been a different sort of individual. But, as the saying goes, if pigs had wings they could fly. In Hill's case, the missing attribute was not a pair of wings, but the diplomacy skills required to navigate academia's sometimes treacherous interpersonal waters. As Taylor put it, "He was rather hot headed according to his own admission." Hill himself later recounted a number of anecdotes about this time. He had been at the university less than a year, for example, when the acting chairman of the faculty summoned Hill to his office to discuss plans to create a new position, president of the university. The chair, who had set his sights on that position, reportedly accused Hill of instead supporting another colleague, a friend of Hill's, for the job. Hill replied that he was not supporting anyone. In Hill's later words, "He called me a liar. I called him a [well, I learned what I called him on the cattle trail]. From that day I knew my tenure of office was doomed. Hardly had his door closed behind me before the policy of pinpricks was begun."

By the end of 1889, Hill's frustrations with university administration reached a boiling point when an executive committee turned down his request for a microscope, even though the regents had approved the request earlier. In a November 1889 letter, John Branner offered commiseration: "I don't think I ever so thoroughly comprehended your surroundings as I did when I contemplated the result of that Faculty vote on laboratory work. We may be in the woods up here," Branner wrote from Arkansas, "but we're not that deeply immersed." In December 1889, Hill submitted a letter of resignation, a gambit on his part, not a statement of a true intent to leave. Called to appear before the board of regents the following February to explain the reasons for his resignation, Hill explained in detail his discontent, which, according to the minutes of the meeting, "were mostly of a personal character." The explanation did not go over well. "Hill," one regent told him, "I'd have bought you a hundred microscopes rather than have had you do that." The board unanimously accepted Hill's resignation and swiftly elected a successor to fill the position.

At this juncture, Hill was counseled by William Dall, an old USGS colleague to whom Hill had complained about mistreatment. Dall began his reply with words of sage wisdom: "No man understands any other man fully, most men fail to understand themselves. So you must not expect to be understood except as far as by tact, prudence, and positive results you may build for yourself a pedestal on which others may put you inside their minds." Dall continued, "I don't think on the other hand you fully understand the generally friendly feeling which geologists here have for you," adding that his colleagues had little time or inclination for "psychological riddles." He concluded, "Talk of what you have succeeded in, and bury your worries and regrets in silence. I never knew a man, however able, who gave other people to think he had hard times (as we all do in this world) that wasn't on the whole the worse off for it in the long run." There is no evidence that Dall's advice was taken to heart by Hill at the time—or ever. In the aftermath of Hill's departure from the university he remained depressed, swallowed up by feelings about perceived mistreatment.

Hill did not return to the USGS immediately, however, instead accepting a position with the Geological Survey of Texas. By August 1891, Hill found himself embroiled in political wars with a new set of colleagues over a new set of issues. He crossed swords with the director, Edwin Dumble, the perhaps aptly named geologist whom Hill himself had recommended for the position not too many years earlier. After Hill joined the same agency, however, the relationship soured quickly. Hill eventually accused Dumble of appropriating his results without proper credit; he moreover stated that Dumble had never made a single original contribution to the knowledge of Texas geology, instead having contributed only "confusion, from which science cannot recover for years." Once again Hill tendered his resignation; once more, to the surprise of nobody familiar with the situation, the resignation was accepted.

The events that played out following Hill's departure from the Geological Survey of Texas provide a quintessential example of Hill's lifelong modus operandi with difficult interpersonal relationships. Hill had scarcely been alone in his feelings about Dumble. Ralph Tarr, a geologist recruited into the Texas Survey by Hill himself, had crossed swords with Dumble more than once. One dispute involved an initial agreement by Tarr to publish one of his papers after another paper, one that Dumble had a stake in, had been published. When publication of the other paper was unexpectedly delayed, Tarr's paper appeared in print first, leaving Dumble furious. Around the same time, while doing fieldwork in the Guadaloupe Mountains, two of Tarr's horses were stolen. On this particular score, Hill's sympathy for Tarr went only so far. In his estimation, the "poor Massachusetts fellow" had ventured into the field wearing a "monkey cap," without "sense enough to look like a Texan." Notwithstanding these feelings, Hill took exception to Dumble's handling of the incident: docking Tarr's final paycheck one hundred dollars, about two weeks' salary, to pay for the horses.

Over the year following Hill's departure from the Texas Survey, a number of other employees resigned, some to take better-paying jobs, but others, including Tarr and Hill as well as a few others, out of discontent. In April 1891, one discontented colleague, J. H. Herndon, brought formal charges against Dumble in a tract titled "Incompetency, Plagiarism, and Maladministration in Office." When informed of the issues, then-governor of Texas Jim Hogg wondered how the man had come to be in his position in the first place. Herndon replied that Dumble had powerful political connections. Hogg went on to explain, however, that neither he nor the state legislature had any control over the state Geological Survey, referring Herndon instead to the state commission office that oversaw the agency. Thus did the head of this commission, a friend and ally of Dumble's, come to be the "judge, jury, and prosecuting attorney" handling the charges.

Going into the hearing, Hill and Herndon felt confident that they had the support of many of their former colleagues, men they knew to share their misgivings about Dumble. Instead, some of these men appeared at the hearing as witnesses to refute the charges against Dumble. The commissioner, inevitably, found the accusations baseless. Hill and Herndon continued to press their case, fearing their own reputations would be damaged if they did not ultimately prevail. Tarr, who had stood with them earlier, backed away from the fight. "Don't you think," he wrote to Hill, "people will begin to believe we are wrong and that we are quarrelsome?" Tarr had come to appreciate, somewhat belatedly, the peril of being on the factually right but politically wrong side of a battle; by the end of 1892, "[Tarr] and D[umble] . . . buried the hatchet." This realization never dawned on Hill himself, nor on Herndon. Hill had stood, perhaps mostly justifiably, on principle, but he had not only taken on a battle he couldn't win, he had doubled down with increasing personal animosity, coming away bruised and, indeed, with a growing reputation for being "hard to get along with."

Echoes of the Dumble affair can be found in many of Hill's battles throughout the years, to the point that authors of a late-twentieth century scientific paper that included a brief retrospective of Hill's career were moved to note with an asterisk "each mention of a position from which Hill was fired or quit after a dispute." Although not entirely lacking in insight and introspection, Robert Hill could embody the dictionary definition of the word "impolitic." He got along well with the people he got along with; with others, not so much. Whether or not he was in the right was besides the point, as William Dall had counseled to no avail: stewing over mistreatment, perceived or real, accomplished nothing good. Quick to take offense, prickly in the extreme about even constructive criticism, and especially quick to see personal motivations behind the actions of others, he sometimes charged into war when the other side had never declared battle. Even when he perceived mistreatment or malfeasance correctly, he took on battles when more prudent colleagues shared his views but had the sense to lie low. What's more, once he sank his teeth into a fight, he really sank his teeth into a fight. Robert Hill didn't just keep grudges; he nursed them, he fed them, he allowed them to drag him down, ultimately accomplishing nothing except damage to himself.

As Powell had predicted, eventually Hill did return to the USGS, given a permanent position in 1892. Powell "received [him] with open arms and a jolly reminder of his previous farewell warning." Hill and Jennie enjoyed a few good years in Washington, DC, entertaining distinguished colleagues, including Powell himself, along with many other scientific dignitaries. On October 1, 1893, Jennie excused herself from her hostess duties, retiring to a bedroom where she gave birth to their daughter, Justina Hamilton Hill. The Hills continued to entertain guests in their home during the years following their daughter's birth. In March 1896 Hill wrote out a list of forty colleagues who "[came] to [his] house one evening for a sociable time." The fact that he made and kept the list hints that, professional successes notwithstanding, Hill never managed to shake insecurities that took root during his formative years, when criticism and restrictions vastly outnumbered encouragement and freedoms. Throughout his life, Hill was especially quick to bridle at slights, real or perceived, from colleagues, including Dumble, whom he had once considered friends. Many years later, daughter Justina alluded to these insecurities as the reason her father had joined a great many professional associations throughout his career. In his own eyes, the man who came of age on the Texas frontier never quite shook the nagging sense that, among his professional milieu, he was not really part of the club.

Hill's return to the US Geological Survey eventually brought him in contact with Bailey Willis. Through the late 1800s, Willis's career in the USGS had moved forward as well, on a smoother track and one that was mostly separate from Hill's. While Hill pieced together the geology of Texas and neighboring states, leaving for Texas and then returning to the Survey, Willis's work had focused on the Appalachian Mountains. In addition to his scientific work, Willis was increasingly tapped by management for administrative roles. Following his stint as editor of *Geologic Maps* from 1893 to 1897, he was tapped to serve as geological assistant to then-director Charles Walcott. When Walcott reorganized the Survey in 1900, he promoted Willis to a new position: chief geologist and chief of the Areal Geology Division. During these years, USGS leadership had an unfailing, passionate champion in Bailey Willis. In a 1904 letter to his wife, for example, Willis described a visit with Walcott to a university during which a professor "pitched into" the work done by two Survey geologists. "I was forced," Willis wrote, "in their absence to defend them and the Survey's methods. [The professor] doesn't love me any more, but Walcott remarked just now that it was what he expected of me."

Although their work as geologists proceeded in mostly parallel non-overlapping orbits, Hill's and Willis's paths did intersect notably on one occasion. At the beginning of 1898, Willis accompanied Hill to the mining region of Llano and Gillespie counties in central Texas. By that time, the USGS had created maps for some parts of the state; the stated purpose of the trip was to identify regions to be mapped next. Almost as openly stated was the young federal agency's hope to convince state leaders to contribute financially to future work. Two decades after the agency's founding, the USGS's financial footing remained far from secure. And three decades after the Civil War, continuing states' rights squabbles reared their head in questions of both financing and control of geological investigations. For its own sake, for political as well as financial reasons, the USGS needed to proceed in cooperation with state agencies. "[Willis] said it would pay the state of Texas," the *El Paso Herald* told its readers, "to cooperate with the government meeting the expenses necessary to give the state a thorough and deserving topographical and geological survey." The article continued, "It will cost the government some $15,000 to survey Texas, and as the state gets all the benefit of this it would seem that the state might contribute something toward the expense."

Hill's pocket diary includes only brief notes about the trip, noting the stops and a couple of times that he advanced small sums of money to Willis. A faithful and prolific correspondent throughout his long life, Willis chronicled the Texas trip in far more detail in letters sent to his mother. On January 3, 1898, Hill met Willis's train at Monett, in southwestern Missouri, where Hill had arrived the day earlier following a visit to Nashville. "With a hearty greeting," Willis wrote to his mother, "he led me into the pleasant breakfast room where we were sumptuously feasted after the fashion of a 'Harvey' eating house. The name is a synonym for good food, prompt service, and cleanliness." The establishment might have agreed with Willis, but the restaurant's other patrons did not. "At my table sat six surveyors apparently long separated from the centers of refinement. . . . Their chief," Willis continued, "was a man of huge body, large . . . beard, and small, very small brain cavity." By all indication, however, Willis found Hill himself to be companionable and the trip itself a congenial adventure. On January 9 Willis wrote from Austin, Texas, that he had planned to spend the morning writing, "but Hill, backed up by a nice horse and buggy and the delicious weather, had me off to Mt. Bonnell and there the beautiful view from under the overhanging cliffs above the blue river charmed me into forgetfulness of the time."

"During the past week," he wrote on January 9, "Hill and I have traveled by train, and buggy and bicycle, 588 miles, all in the prairies of eastern Texas." They visited Round Mountain, which Willis described as a little hill. "Hill pointed out several plants like the melon cactus and a shrub, the agarita, which do not grow in the surrounding prairie or elsewhere so far north." Willis's series of letters continued to paint a portrait of a pleasant journey. On January 12, 1898, Willis described a bicycle trip the team had taken: "Oh! The rocks and cactus! Hill broke his saddle, and the General lost two spokes and got out of wind." Searching Willis's letters for clues as to who "the General" might have been, one finds no mention of a third party, let alone a general of any stripe. Throughout his life, Willis was known to refer to himself, in various creative ways, in the third person.

Nothing in Willis's letters to Cornelia hints at feelings of ill will toward Hill nor at any sense of discontent with the trip. At the end of their planned travels, the two men even took off on the nineteenth-century equivalent of an impromptu road trip after they discovered they could travel 250 miles southward into Mexico for just a dollar and a half. Of their time in Texas, Willis relayed a series of upbeat reports. "The days are full of interesting work," he wrote on January 11. In the same letter he described an outing to Pilot Knob, an old volcano: "It is not of scenic interest but a geologist find[s] the problems of its former activity exceedingly attractive." In another letter, however, Willis wrote, "Last evening Hill took me visiting. First to see Mrs. Pease, the widow of Ex-Governor

Pease . . . and later with Major Swain a 'carpetbagger from Illinois.'. . . The people of the State," Willis continued, "are very interesting, and there are many classes. It requires an effort to overcome our prejudices and view them fairly sometimes and they without doubt are often incapable of doing so toward northern people. A congressman who sent his sons to be educated at Harvard and Yale endangered his reelection."

For their part, local newspapers trumpeted Willis's arrival in Texas: "There is no higher authority in the United States on all geological matters," proclaimed an article in the *Houston Daily Post*, "than Mr. Willis, and no man's opinion in this respect is worth more than this. . . . Knowing this," the article explained, "the Post correspondent sought an interview with the eminent geologist." An earlier article in the *Houston Daily Post* had noted that Willis was the son of N. P. Willis, the poet, and was "quite a young man to hold a position of such importance as the head of the geological division of the United States geological survey." The visit was of interest to *Post* readers because one of the team's main objectives was to map out the region's potentially valuable mineral reserves, including not only granite and other ornamental stones but also iron and gold ore. The visit garnered wider attention when Willis voiced surprise at the richness of the district in minerals. "He intimated," a *Washington Times* reporter wrote, "that if proper research was made for gold, a veritable Klondike might be developed." This intimation would prove to be overly optimistic; although the Llano region eventually produced good amounts of placer gold, neither the Llano region nor anywhere else in Texas would ever challenge the Klondike's legendary gold production.

From Llano, Willis and Hill headed south to San Antonio, where they planned to investigate the artesian water resources of that city. On January 18, Willis wrote, "This morning we are engaged with those who wish to talk to the 'eminent geologist,' or whom Hill wishes to introduce.

Ex-governor Hogg, for example, and capitalists, etc." Willis and Hill would have been solidly united behind the goals of the trip, namely, to lay the political, financial, and scientific groundwork for further field investigations in Texas. Willis offered up his optimistic talk of undiscovered gold reserves toward this end as a way to stir up interest and support from the public. Meeting with the state's movers and shakers would have been another important means toward the same end. In this regard, Willis made use of Hill's extensive network of local connections. Indeed, though Willis was the appointed USGS emissary and the higher-ranking scientist, he was also a Yankee behind what some still regarded as enemy lines. It made all the difference throughout the trip to have Hill, with his deep roots in Texas soil, at his side.

In the eyes of the *Houston Post*, there might have been no better authority on Texas geology than Bailey Willis, but of course it was Hill who was by then well on his way to being recognized as the father of Texas geology. He had spent years developing a comprehensive report of the Black and Grand Prairies, a swath stretching from northeast Texas through Austin toward San Antonio. Whereas Willis spoke of future riches from gold, Hill had long seen great potential wealth in the same "black waxy" soils that had vexed his first visit to Comanche back in 1874 and in the state's artesian waters. By 1900, Hill would complete the first draft of a monograph initially titled "Geology and Artesian Wells of the Black and Grand Prairies of Texas," in which he presented exhaustive accounts of the geology, geography, and artesian water of the region.

Even before Hill's report was published in 1901, Willis highlighted Hill's work with generous words in an article, "Work of the U.S. Geological Survey," published in *Science* magazine in August 1899. "The geology of Texas," Willis wrote, "is associated with the name of Robert T. Hill. Under his direction a large map has been prepared of the State on the scale of 25 miles to the inch, including portions of Oklahoma, and an account of the physical geography has been written to accompany it." With these words, Willis took pains to give credit where credit was due; it remains unclear why the *Houston Post* had not done so, but it appears the articles written about the trip left a bitter taste in Hill's mouth. Throughout his life Hill kept a scrapbook of newspaper articles in which he was quoted, collected from far and wide; it includes no articles from his trip to Texas with Willis, even though Hill's name did appear in all of the articles as well. Had Hill been irked by local reporters' lapdog attentions to "the eminent geologist"? Or had Willis's bold talk of gold strikes in Texas rankled Hill's more staid—and better-informed—sensibilities and his own (correct) inclination to look toward less flashy potential riches, namely water, fertile soil, and fossil fuels?

Reading Willis's letters to his mother, one wonders about another— or an additional—possibility. Had Willis shared with Hill his thoughts about meeting people whom Willis had to make an effort to overcome prejudices, not realizing—or realizing and not caring—that his colleague counted himself, and proudly, among those same people? Had Willis shared with Hill the contempt that he felt for the rough-hewn surveyors they encountered at breakfast in Monett? The surveyors were not Southerners but, rather, hailed from another group for whom Willis had little respect: the uneducated and (in his estimation) uncultured. Willis might have been oblivious to the fact that his Cornell-educated, scientifically astute colleague did not share his biases toward either Southerners or the uneducated. But indeed, neither bias was shared. As Hill would later write in a soul-searching 1902 letter to a boyhood friend, "I prefer the company and association of plain and practical people and the less a man has his manhood rubbed off by the artificiality of education the more I gravitate toward him. In other words, I do not give a darn for the finished and successful, but only care for those who are either struggling upward or are simply plain people."

Following the 1898 trip to Texas, Willis's and Hill's geological investigations continued to focus on different parts of the country. The 1900 reorganization of the agency, however, put their orbits on a collision course. Whereas Hill had previously reported to Walcott directly, following the reorganization he reported to Willis. In Hill's later words, Willis "had no knowledge of my work or problems." In a 1906 letter to John Branner, Hill put it more bluntly: "Willis, the mamby-pamby was absolutely worthless as a substitute director." Hill's monograph on his work in Texas, submitted in 1900 for publication by the US Geological Survey, was reviewed by three colleagues, among them Bailey Willis. The review committee agreed that the publication would become a standard work of reference for a long time to come and, in light of its importance, required substantial editing to improve the writing. They objected to the proposed title, which they saw as too narrow, proposing instead, "The Cretaceous Formations of Texas: Their Geologic and Topographic Relations, with Special Reference to Artesian Waters of the Black and Grand Prairies." They further found fault with Hill's writing style, which they saw as repetitive and overly studded with adjectives. In Hill's eyes, the unqualified committee had picked nits with the intent of suppressing publication. Pressing forward, he resubmitted a revised report in 1901, with an expanded but different title than the one suggested: "Geography and Geology of the Black and Grand Prairies, Texas, with Detailed Descriptions of the Cretaceous Formations and Special Reference to Artesian Waters." It was accepted for publication as part 7 of the *21st Annual Report of the USGS*. To Hill's further consternation, distribution of the report, however, was delayed in the bindery, emerging for distribution more than six months later.

The monograph that finally did appear was encyclopedic, consisting of 666 pages illustrated by seventy-one plates, six of which were large maps. It was the most comprehensive geologic report that had been written to date on Texas, summarizing all that was known about the geology and geography of the state, including its artesian water reserves. It remains today an invaluable reference for geologists, with original copies fetching several hundred dollars (when they can be found). At the time, it provided a blueprint for further economic development. More than a third of the monograph focused on water: the principles of underground water, the artesian-well system of Texas, and artesian well reserves of the Black and Grand Prairies. Hill's map shows the locations and depth of existing artesian wells, with logs and detailed stratigraphy of many individual wells included. This information was, Hill knew, of vital importance for a growing region, including the expanding city of Dallas, already suffering from want of adequate water supplies. In the work that he regarded as the crowning glory of his career, Hill had also taken care to present technical information clearly enough that it could be understood by nonspecialists, not just other geologists. Following publication of the report, Ellis Shuler, an accomplished geologist who went on to be the dean of the graduate school at Southern Methodist University, described Hill as being "the single largest contributor to the uncovering of wealth of Texas." Shuler is among the colleagues who will make a number of appearances in our story, in his case by virtue of having been Hill's closest lifelong friend in professional circles. Hill and Shuler corresponded extensively with one another throughout Hill's life. Shuler moreover appears to have brought out Hill's better angels; Hill's letters to Shuler included some thoughtful personal introspection and little of the aggrieved tone that sometimes shines through his letters to other colleagues.

Robert Hill as a young man (undated). (DeGolyer Library, used with permission)

In retrospect, as was the case with so many of the battles that Hill fought throughout his career, the truth behind Hill's struggles to get his monograph published remains unclear. Was the volume well written because of, or in spite of, the pains that had been taken with the review process? Had its publication and distribution been delayed in the interests of making sure this seminal reference was as good as it could be, or had dark forces of bureaucratic evil—or personal bias—been somehow at play? Had Hill been overly prickly about valid criticisms or justifiably perturbed? In the end, the process produced a publication heralded for having been especially well written, penned by a man who had been a newspaperman before he was a geologist. It is possible the early draft called out for heavy-handed editing. It is also possible that, in laboring to create a report that a nonspecialist could understand, Hill had run up against the editorial sensibilities of colleagues who were used to a different, more formal style of technical writing. Conceivably the most onerous edits had not come from Willis, the son of a poet, whose own writing could also run toward more colorful, even flowery expression than standard scientific prose. Hill's bristly reaction might have focused unfairly on the member of the editorial committee with whom previous interactions had not gone well. Later in his life, when Hill looked back at his tenure with the USGS, he repeatedly pinned his difficulties on a handful of individuals who he perceived as having been motivated by bias. A 1928 letter, for example, referred to "the land where I had to suffer from the meddlesomeness of Willis, Gannett, and others of that ilk without local support." While scarcely the only USGS colleague with whom Hill had difficulties, Willis's name always appears among those individuals mentioned as having been irksome when Hill did name names.

Wherever the truth lay, the monograph drove a further wedge between Hill and Willis, but it also further established Hill's standing as the leading expert on Texas geology. Over the years that followed, he went on to publish not only technical articles in scientific journals but also popular articles, for which he was paid, by leading magazines. And yet, even at the pinnacle of his career, he found himself discontent. With all of his success, he wrote in 1902, "I am a very lonely and sometimes unhappy man, however. I can not assume professorial dignity, I can not . . . enjoy the company of men solely because they are alleged to be intelligent." By the age of forty-four he recognized in himself a certain pugnacious quality. "I have," he wrote, "an unhappy faculty of taking the opposite side and getting a rise out of people, nine out of ten of whom do not understand that I am merely quizzing them." In the eyes of others, even colleagues who regarded him with esteem and affection, Robert T. Hill suffered a regrettable lack of tact.

Decades later, toward the end of his life, would Hill look back at the painful years of his early childhood—the years of "noxious prohibitions" when there was never "a caress or a pat or a friendly word of approval." Whereas Bailey Willis's childhood had been happily lonely, his sails buoyed by the inexhaustible wellspring of his mother's love, Hill's childhood had just been lonely, with few if any happy winds to fill a young boy's sails. Looking back, Hill concluded that his childhood mind "became as full of buried complexes and repressions as a poor man's dog is full of fleas." He later wrote to an old friend, "That is why I have (an irrational?) inferiority complex and delusions of persecution."

Looking back, one wonders if there were reasons apart from childhood experiences that contributed to Hill's prickly tendencies, his sense of inferiority. In modern halls of academia, some—usually women and underrepresented minorities—sometimes speak of "imposter syndrome": a nagging sense, which professional accomplishment cannot easily dispel, that one is not really good enough. It has been suggested that imposter syndrome is not a consequence of innate insecurity but, rather, a natural consequence of being on the receiving end of subtle and not-so-subtle messages that one is, in fact, not really good enough—that one does not, in fact, belong. More generally, the problem with being treated unfairly some of the time is that one can start to perceive injustice behind all of life's frustrations, facing the world with a sense of grievance that takes on a life of its own. An appealing sense of confidence, perhaps edging toward overconfidence, and good humor comes naturally to the man who has been overwhelmingly treated fairly in life, a man like Bailey Willis. To an individual who has faced hardship and real injustice throughout life, bubbly confidence can be harder to come by.

Robert Hill, along with essentially all of his contemporaries in geology, hailed from the majority demographic of the day: white, male, of western European descent. Yet, as he first realized in the "mamby-pamby" hygiene lecture at Cornell, within his professional milieu he was part of an underrepresented minority, a (by then) rough-edged boy who had clawed his way to higher education. He was a boy who had come of age on the Texas frontier, unlike the contemporaries with whom he found himself, boys like Willis, groomed by privilege to be part of an intellectual elite. He was, moreover, a Southerner who had seen and felt the Civil War firsthand, raised by a grandmother with abiding enmity for Northerners, shaped as a young adult by life on the Texas frontier. Over the years, indeed, decades, following the war, many in the North saw no need to hide their feelings about Southerners. Hill recounted an experience during his years at Cornell when the "Iowa lady" running the boardinghouse in which he resided for a time, not knowing that he himself hailed from Tennessee and Texas, "announced that if she knew there was [a Southerner] in the house, she would ask them to leave." (This one anecdote as much as anything else suggests that, in adulthood, Hill did not speak with an identifiable Southern accent.) At least some of the prejudice he perceived, starting with the stern elderly teachers sent into the South after the war and continuing at Cornell and undoubtedly to some extent during his years in Washington, DC, had not been merely a figment of Hill's imagination.

Hill would later refer to Bailey Willis as a "needlessly offensive man" who "hated all things Southern." Was this characterization a figment of Hill's imagination? During the 1898 trip to Texas, Willis spoke of the difficulty of overcoming prejudices, on both sides. Other letters provide additional direct glimpses into how Willis regarded Southerners. In 1890, years before his trip to Texas, Willis traveled to Knoxville with his first wife, Altona, and their daughter Hope. He wrote to his mother that during the trip, he "called and found Mrs. Sanford to be a most charming

French lady, speaking English with a pleasant accent although she has been many years in this country." Mrs. Sanford, who was actually Swiss by birth, was part of the Knoxville elite; her son, Edward Terry Sanford, went on to serve on the US Supreme Court. Willis's letter continued, "These are the first people of real culture we have met in the South." It was not as if Willis had had limited experience in the South; by 1890 he had undertaken extensive work on Appalachian geology, sometimes establishing a home base there for months at a time. Letters he wrote during this time reveal that he did not wear his prejudices on his sleeve, so to speak. Or at least he did not dwell on them. Throughout his life, his letters to various family members were upbeat, with only an occasional unvarnished comment about a disagreeable circumstance or character. His remark about his visit with the Sanfords suggests, however, that Willis's appreciation of culture might indeed have stopped at the Mason-Dixon Line.

Another letter, written by Willis on October 30, 1898, describes his first encounter with John Branner, who by that time had left Arkansas to become the chairman of the geology department at Stanford University. Willis's account of the meeting begins with an explanation of Branner's lineage, including the fact that his grandfather had moved in the eighteenth century from the "valley of Virginia" to Tennessee. A man's lineage was, it seems, an important matter to Willis. Although he was not baldly critical, one's ear detects hints of condescension toward a colleague who had held a number of auspicious positions, who by 1898 had been plucked from Arkansas to lead the new geology department at Stanford. "Work," Willis wrote, "was the idea of Branner's education, thrift is strongly ingrained, and common sense is his chief mental characteristic. He is slow at ideas, but positively sure." Bailey Willis was, like any man, but some more than others, a product of his upbringing. He was not only a Yankee to the core, but also part of a Northern intellectual elite that did not hesitate to judge biases in others, but could perhaps be myopic where their own biases were concerned.

Hill's upbringing, on the other hand, had shaped a very different character. The experiences of Hill's childhood and early adulthood toughened his hide, in a way similar to but different from Bailey Willis's experiences during his teen years with school-yard bullies and later experiences with rough-edged woodsmen. Whereas Willis met the challenges of early adulthood with the confidence of a young man who had been nurtured and adored throughout childhood, Hill's sense of self-worth had been battered far more than it was ever nurtured. It would have been surprising if the experiences of his formative years had not left a mark.

Also a product of his upbringing, Hill found a natural nemesis in Charles White, in the late 1800s also part of an aristocratic intellectual elite, and, later, Bailey Willis. History shows that both Hill and Willis found themselves frustrated with bureaucracy through their early years with the USGS. Although they were almost exactly the same age, Willis was tapped for management roles that, while prestigious, were burdensome to an energetic young geologist. In contrast, Survey managers, including Powell himself, appreciated his talent as a geologist, but did not see Hill as management material, which freed him to focus on geological investigations but often left him to rail against management. One of these battles, in the years following publication of his great monograph, led him to write, "It is not the barking of the jackals that hurts me so much, as the habit they have of getting in one's way and temporarily obstructing the course of life."

Viewed through the lens of hindsight, one cannot but be struck by how much Hill had in common with Willis. Whatever manhood had been rubbed off of Willis by his prep-school education, he had earned at least some of it back in the snowy north woods of Minnesota and the rugged terrain of the Pacific Northwest. It is tempting to think that, if only the two men had spent more time together, they would have discovered common ground. Yet they did spend time together; their 1898 trip to Texas put them in each other's company for an extended journey, the kind of experience that is known to draw individuals closer. The experience appears, however, to have not been a profound bonding experience—and, viewed from Hill's eyes, perhaps understandably so. The biases that Willis revealed with his own words would have been irksome to Hill, but not only that. Although even at the time he had been recognized for his pioneering investigations in Texas, it was Willis whom the *Houston Post* proclaimed to be the most knowledgeable authority in the country on all matters geological, including the geology of Texas. And whereas Hill focused on the vital importance of mapping artesian water reserves, Willis made headlines in Washington, DC, newspapers with buoyant hints that Klondike-scale gold reserves might be found in Texas. Both fundamentally made the same argument: that past and future geological investigations were critical to the state's economic development. But they could scarcely have come to the argument from more different places, or with more different styles. Whether these two men agreed on something or not was beside the point; fate had not only decreed that their professional paths would sometimes cross, but also that, when this happened, they would view the world from very different perspectives.

On the occasion of the 1898 trip, united by a shared sense of purpose, Hill might have been content to swallow, or at least stay quiet about, his reservations about perceived grandstanding on Willis's part. But Willis's sensibilities as well as his modus operandi had been on full display throughout the trip, and Robert Hill had a front-row seat to his colleague's antics. If Hill never said much about those antics at the time, they cannot have but shaped his views of his colleague. Those views would resurface many years later, when circumstances again drew the two men's orbits into a collision course.

CHAPTER 4

PARTING COMPANY
AND FACING DISASTER

Life is either a daring adventure or nothing.
—*Helen Keller*

By any objective measure, both Hill's and Willis's careers with the US Geological Survey were, if occasionally bumpy, highly successful through the 1880s and 1890s. Hill's frustrations had certainly resurfaced and redoubled after an agency reorganization left him reporting not to the director, but to a man for whom Hill had what biographer Nancy Alexander called a "pungent contempt." Among the incidents that stoked Hill's feelings was an instance when Willis "ransacked [his] office and papers" during his absence to make "what [Hill] considered a false report." Once again Hill saw old prejudice at play. Reading the words that Willis himself wrote, during his 1898 trip with Hill to Texas and at other times, it seems Hill might have come by this assessment to some degree honestly. As the joke goes, just because you're paranoid, it doesn't mean they aren't out to get you.

Yet Hill persevered at the USGS, working during summers with great personal freedom in Texas and during winters in the Caribbean under the direction of Professor Alexander Agassiz of the Museum of Comparative Zoology at Harvard University.

Hill's work in Texas focused increasingly on the Trans-Pecos region, stretching between the Pecos River and the Rio Grande in Texas, extending into eastern New Mexico. This then-virgin part of the state had been described as a "Land of Despair," where a traveler had to "climb for water and dig for wood"—references to, on the one hand, the need to sometimes climb windmill towers to turn the wheel by hand and, on the other hand, to dig for mesquite roots. Hill had first explored the area back in 1889 with none other than John Wesley Powell. A decade later, then-director Charles Walcott authorized Hill to lead an expedition down the Rio Grande. Since 1845 the river had marked the international boundary, but an official surveying party failed to get through the canyons in 1852, deeming them impassable. For the next half century, the exact shape of the United States' border with Mexico remained unmapped. By 1899, Hill's acclaim in a public as well as a scientific arena was such that the *New York Times* published a small article about the voyage, describing Hill as "bold and adventurous" as well as the "ablest, probably, of the younger American economic geologists."

For his assault on the canyons, Hill led a five-man team including his nineteen-year old nephew, a boatman, a cook, and two frontiersmen. One of the frontiersmen, James McMahon, was the only man known to have survived the trip previously—not for an official expedition, but in search of beaver pelts. Hill's journey was arduous almost from the start, with concerns about a notorious local bandit by name of Alvarado adding to the natural challenges, of which there were no shortage. Hill's team stopped many times each day to carry their three boats around rapids deemed too dangerous to ride out. Each such excursion was wearying, but paled in comparison to the challenge that Hill's team faced in a place

Photograph of Gran Cañon de Santa Helena along the Rio Grande, dubbed Camp
Misery by Robert Hill's field team, taken during Hill's first expedition of that river.
(USGS photograph)

known as the Gran Cañon de Santa Helena (sometimes called Santa Elena), where toppled cliff rock had turned the river into a jagged 200-foot-high (about 60 m) obstacle course. Over the course of three days at a place they named Camp Misery, they carried the contents of their boats over the rocks before facing the most daunting challenge: lifting the 300-pound (135 kg) boats over the "vast cubes of limestone." Hill and his team spent their nights wedged into "such perches as we could obtain upon the short-cut edges of the fallen limestone blocks, above danger of flood." Even in the midst of physical tribulation, the splendor of the setting was not lost on Hill: "We had abundant opportunity," he later wrote, "to observe the majestic features of the great gorge in which we were entombed." The entombment happily proved temporary.

Although Hill and his team were not always in the best spirits along the way, they completed the 350-mile (about 560 km) journey, the first successful scientific expedition down the river. With the observations he made, Hill put together a first-ever geologic section along a key geologic corridor, "procuring light upon some of our least-known country." He spent the next two field seasons investigating the Trans-Pecos region and began to formulate plans for a comprehensive technical report. In 1900 Hill wrote what Bailey Willis later called "a picturesque account" of the expedition for *The Century Illustrated Monthly Magazine*, a widely read and well-respected popular magazine.

Among Hill's professional accomplishments during these same years were seminal investigations of the geology of Puerto Rico, Haiti, and Cuba. On May 9, 1902, literally as Hill worked on a manuscript on the Antilles Islands, the first media reports reached US shores describing a May 8 eruption of Mount Pelée that had annihilated the city of Saint-Pierre, Martinique, formerly a jewel of the Caribbean. A follow-up media story, including a report by the first mate of one of the few ships anchored offshore that had not been consumed by the disaster, soon provided grim verification of the horrific early reports. The first mate described the scene: "There came," he wrote, a "sort of whirlwind of steam, boiling mud and fire, which suddenly swept the city and the roadstead" and barreled forward to consume the harbor as well. "There were some eighteen vessels anchored," he added. "All of the vessels immediately canted over and began to burn." All the ships except the *Roraima* "sank instantly and at the same moment." Scientists would later estimate the heat of the blast by considering which things had melted (glass but not copper) and which things had caught fire (wood ships and human skin). Other newspaper articles included more lurid details, describing "Thirty thousand corpses . . . strewn about, buried in the ruins of Saint-Pierre, or else floating, gnawed by sharks, in the surrounding scene." Twenty-eight "charred, half-dead human beings" had been brought to Fort-de-France, of whom only four were expected to recover.

The US media rushed to interview scientists about the disaster—Hill, by then a leading expert on the geology of the West Indies, chief among them. Asked by a journalist to describe the cause of the disaster, Hill explained what was known about volcanoes at the time, adding that volcanism was "still one of the most inexplicable and profound problems to explain." He went on to observe that volcanic activity was known to sometimes break out "in widely distant portions of the earth." He pointed to recent eruptions in Colima as well as Chilpancingo, Mexico, as well as tremendous earthquakes that had recently struck Guatemala. Scientists today understand that volcanic activity and earthquakes are sometimes linked; volcanic eruptions are often accompanied and/or preceded by local earthquakes, and large earthquakes can sometimes trigger unrest in nearby volcanoes. It remains highly controversial at best, whether volcanic activity does in fact break out in widely distant regions as a consequence of a physical linkage. Scientists today overwhelmingly discount links between distant volcanic centers. But in 1902, geologists still lacked the basic physical framework of plate tectonics, which half a century later would provide the overarching explanation for volcanoes as well as earthquakes. Hill knew, and explained, that volcanoes were not understood.

On May 13, 1902, the National Geographic Society commissioned Hill to represent the society on a relief expedition to Martinique. Consequently, Hill was among the first scientists on the scene, reaching the epicenter of devastation on May 22. For the scientists and reporters sailing toward St. Pierre on a tug, the first view of the "ghastly, ashen-gray" landscape brought a "cry of horror from every person on board." That sense of horror was about to reach Hill on a far more personal level. He chartered a steamer and made frequent landings to survey the scene, mapping the zone of destruction. Not content with even these forays, he returned to Fort-de-France, looking to arrange horses and supplies to explore farther up the slopes of the still-smoldering volcano. Terrified and traumatized natives boggled at the idea that anyone would go toward the scene of devastation they were so desperate to flee.

By May 24, Hill had located two animals and two men, an army soldier and a young interpreter. They set off on the main road northward, Route de la Trace. To Hill's surprise, the first part of the journey revealed no hint of devastation but, rather, only the beautiful, lush scenery that still today characterizes much of southern Martinique. Hill developed a nearly instant distaste for the soldier, however, who looted every house they passed; his affinity for unfinished men did not extend to dishonest ones. The two soon parted company. Newspapers back home, including the *Fort Worth Morning Register*, described Hill's "daring and prolonged investigation. Prof. Hill," the article explained, "is the first and only man who has set foot in the area of the craters, fissures, and fumaroles." The

UNCLE SAM'S MEN FIRST IN SCIENTIFIC RESEARCH AS WELL AS IN RELIEF WORK.

Cartoon accompanying one of the many newspaper articles written about Robert Hill's 1902 investigation of Mount Pelée. (*Pittsburgh Post*, May 29, 1902)

article had been written as Hill had prepared to set out on horseback to make his way even farther up the volcano to reach its central craters. "The undertaking is very hazardous, as eruptions may occur at any moment," the article continued. "Prof. Hill knows the risks he takes, but says the only way to discover exactly what has happened is to go to the crater itself, or as near it as possible."

Fort Worth readers could not have imagined that, by the time they read this article on May 28, Hill's appreciation for the risks had become exponentially more acute. Toward the end of the day on May 26, Hill arrived at the small village of Fond St. Denis, which sits just 3 miles (about 5 km) below the summit of Pelée. At the time Fond St. Denis was a hardscrabble collection of buildings strewn out along the road that snaked through a narrow canyon toward Saint-Pierre. After passing a small church then under construction, Hill met a small group of people, "one of whom kindly placed at our disposal an empty cottage" where they could spend the night. As twilight descended, the sky above the volcano was suddenly lit by a "dim flare of light like the sheet lightning of a summer storm." In another article, Hill wrote, "Following the salvos of detonations from the mountain, gigantic mushroom-shaped columns of smoke and cinders ascended into the clear, star-lit sky and then spread, in a vast black sheet, to the south and directly over my head. Through this sheet, which extended a distance of 10 miles from the crater, vivid and awful lightning-like bolts flashed with alarming frequency." Watching the growing cloud with horror, Hill realized that, while it seemed to spread slowly, it moved "so fast that it was easy for me to see that it was not to be escaped by running." From a vantage point far too close for comfort, Hill became the first scientist to personally witness and describe what we now know as a Plinean eruption, a style of eruption that blasts great clouds of gas and debris into the atmosphere.

Knowing he could not outrun the advancing cloud, Hill squatted next to the building, continuing to take notes and photograph the continuing series of flashes. He would later explain that he had "bungled" the exposure, capturing only a single flash on film; one imagines the situation would not have inspired the best work in any photographer. As the cloud neared Hill at 8:40 p.m., he placed his notes under a jug on the ground in the hopes that they, at least, would survive. Though he faced apparently certain peril, fate intervened. "Mysteriously and silently," Hill later wrote, "with a change of wind, the cloud shifted its course and moved toward the east." For a few days, newspaper articles reported that Hill, the "noted explorer," was missing, having set off two days earlier to explore the mountain, "the sides of which [were] still red-hot from the flow of molten matter." An article published on May 28 added, "Friends of the noted explorer are beginning to worry."

Spared by the last-minute change in winds, Hill had in fact fetched up in Fort-de-France on the evening of May 28, "completely worn out" by his trip: "My attempt to examine the crater of Mont Pelée," he wrote, "has been futile." He went on to explain that, from this vantage point, "near the ruins of St. Pierre, [he witnessed] a frightful explosion from Mont Pelée, and noted the accompanying phenomena. While these eruptions continue," he added, "no sane man should attempt to ascend to the crater." Explaining that he had taken many photographs, he further added, "but [I] do not hesitate to acknowledge that I was terrified."

Ironically, Hill's quest to understand Mount Pelée led him and some of his American colleagues astray scientifically. Hill concluded, not unreasonably, that the eruption he had witnessed provided the key to understanding the eruption that had destroyed Saint-Pierre on May 2—that is, that a massive column had risen into the air and then collapsed under its own weight, overcoming all in its path. Lacking proper specimen collection bags, he contributed his own socks to bring back ash for further analysis. "Our greatest problem now," he said, "is to determine the nature of the gas which sank down on the city of St. Pierre." Drawing on geological observations and eyewitness accounts, French geologist Alfred Lacroix later concluded that the destruction had been caused not by a collapsing cloud of gas but, rather, by a lateral blast, what scientists now call a pyroclastic flow. It fell to Lacroix to first describe in detail what he called the "*nuée ardente*" (glowing cloud) that had overcome Saint-Pierre. With temperatures reaching more than 2,000 degrees Fahrenheit (1,000 degrees C) and speeds upward of 60 miles (97 km) per hour, a pyroclastic flow can be very lethal, very fast.

Although Hill's stature as a geologist had been established by his work in Texas, and his conclusions about the 1902 eruption were more wrong than right, his exploits in Martinique, described in articles in newspapers around the world with headlines such as "HE DARES DEATH TO AID SCIENCE," elevated to new heights his fame and renown among the public. Even Tennessee stepped forward to claim Hill, "a man of wonderful vigor and indomitable courage," as a native son. Hill later filled a 200-page scrapbook with newspaper articles published around the country about his exploits.

Even before his harrowing firsthand experience with the explosive forces of volcanism, the eruption of Mount Pelée left a mark on Hill's sensibilities as a scientist. "When I endeavor to write a scientific description of the volcanic disaster in Martinique," he later wrote, "I cannot keep my statements dissociated from the human aspects of the subject. Mingled with every scientific proposition there is a thought of human ruin and disaster, a swirling of thoughts of the geologic causes and corpse-strewn streets."

From the start of his career, Hill, like Willis, had been concerned about the landscape: sorting out the large-scale geological puzzle pieces. There is little evidence that either of the two geologists paid much attention to earthquakes. Willis's native New York differenced from Hill's native Nashville and adopted state of Texas in many ways, but the three places are not entirely unalike in terms of earthquake hazard. Earthquakes do occur in all three regions, but far less frequently than in California and other parts of the West. Both men would have been aware of the earthquake that struck Charleston, South Carolina, in 1886, an earthquake that caused heavy damage in Charleston and generated perceptible shaking as far away as the Mississippi River valley. But neither man was involved with investigation of that event. In his early experiments with wax models,

Willis, an engineer by training, had shown more interest than Hill in understanding how the landscape had evolved over time into its present form. Both men were, however, first and foremost field geologists at heart.

The organization that both Hill and Willis worked for at the start of their careers, moreover, had been created, quite literally, to survey the geology of the United States. In modern times the USGS is known for its preeminence in earthquake monitoring and hazard assessment, but the agency's focus on natural hazards only began in earnest nearly a century after its founding in 1879, when then-President Jimmy Carter first signed legislation in 1977 to create the National Earthquake Hazards Reduction Program. Before the creation of NEHRP ("knee-herp," as it is known), the USGS had had some limited involvement with earthquakes. As early as 1886, the Survey took a leading role in the investigation of the Charleston, South Carolina, earthquake. Captain Clarence Dutton, another nineteenth-century geologist pressed into service during the Civil War, had joined Survey ranks in its earliest days and led one of the most comprehensive investigations that had been undertaken to date of an earthquake. The so-called Dutton report, published in 1889, not only included an exhaustive characterization of the effects of the earthquake, but also presented a pioneering use of the recorded times of earthquake waves at different locations to estimate the speed of seismic waves in the earth. Two decades later, Myron Fuller, also with the USGS, published the first comprehensive geological investigation of the New Madrid earthquakes, which had struck the boot-heel region of Missouri back in 1811–12.

Perhaps inevitably, an agency created to understand geology concerned itself from almost the start with occasional geological paroxysms, earthquakes as well as volcanoes. As plate tectonics theory emerged, geology itself, or at least big branches of the field, became a more dynamic science. The nation's preeminent geological agency evolved as well.

Neither Robert Hill nor Bailey Willis lived to see the advent of plate tectonics theory or the later expansion of the USGS mission; they died in 1941 and 1949, respectively. Both men's talents lay in classic field geology, which they put to work in one part of the world or another through most of their long careers. Yet it's hard to spend one's professional life in any earth science discipline without developing some appreciation of and interest in the ways that geological forces can sometimes menace human existence. For Hill, the eruption of Pelée was an indelible experience, even before his harrowing experience on its flanks. Before he left the safe confines of Washington, DC, for Martinique, he pointed to the disaster to underscore a point he had made earlier, when he found himself in a position to weigh in on the location of a proposed transcontinental canal through Central America. By then, Hill had spent some years investigating the geology and structural setting across Panama and Costa Rica, work that he had published in another comprehensive report. In an 1896 article written for *National Geographic* magazine, Hill described the advantages of three proposed routes: a ship-railway corridor across Mexico or a canal across either Nicaragua or Panama. Panama, Hill concluded, offered not only the most economical route but also a safer option than Nicaragua, where destructive earthquakes occur more frequently. Hill's expert opinion did not settle the matter immediately; when Mount Pelée erupted in 1902, a canal across Nicaragua remained under consideration. By the following year, the Panama Canal Treaty was ratified by the United States and Panama, giving the United States ownership of a path extending 5 miles (8 km) on each side of the proposed canal, in exchange for a yearly payment of ten million dollars.

The year 1902 might have brought Hill to the pinnacle of his renown and redoubled his appreciation for natural hazards, but it marked the beginning of an unhappy decade in his life. The eruption of Pelée had elevated Hill's public fame to new heights but, to the growing consternation of USGS management, it derailed progress on the important Trans-Pecos report. The consternation was a two-way street, with Hill continuing to chafe within an organization whose structure left him subordinate to Willis.

While still employed with the Survey, Hill entered into a partnership with Joseph Qualey, a businessman who proposed to develop mining enterprises in Mexico with Hill's expertise. Hill lived a draining double life, shuttling between his official duties in Washington and a business office in New York. Then the relationship with Qualey began to turn south. In early 1903, Hill concluded that Qualey was using Hill's reports and stature to raise funds for himself, with no intention of paying Hill for his services. He had entered into the business relationship, Hill wrote, only to make money that would allow him to better pursue his interests in science—and found himself swindled. The year proved bleak in other respects as well. By this time his family was in a state of collapse, his mother and one of his brothers having died and his marriage in shambles. Jennie Justina, who had captivated Hill at Cornell, had often joined her husband on his travels early in their marriage, but over time spent less and less time with him , especially after the 1893 birth of their daughter Justina. Jennie retreated increasingly to her family's home in Massachusetts, leaving Hill "miserable and lonely" in Washington.

In the midst of Hill's despondency, he once again found himself crossing swords with Survey management regarding the overdue Trans-Pecos report. In June 1903, the USGS mailed Hill an official order, notifying him that his salary was being reduced to $1,000, to be paid upon completion of the report. In Hill's eyes, it was an outrage, the "last degradation that [he] could stand." He wrote his letter of resignation, dated to take effect the following day. Contentious dialogue ensued between Hill and USGS management, including, eventually, director Charles Walcott. In reply to Hill's resignation letter, his then-supervisor expressed surprise as well as concern about the Trans-Pecos report; Walcott's letter repeated that $1,000 had been set aside for payment to Hill upon completion of the report. In the spring of 1904, Walcott again extended the same offer. Hill sent a peevish reply, pointing to Willis specifically as the cause of his difficulties: "I had five years," he wrote, "of Mr. Willis' intervention and interruption of my scientific research and work and it is simply a physical impossibility for me to contemplate a repetition of these conditions." He deigned to conditionally accept the facilities of the USGS to help him complete the report, but refused the money, "as I have done all of my scientific work for the love of it."

His letter to Walcott raised the possibility that he had been treated unfairly because of "the well-known fact that I am a Southern Democrat; my failure to participate in scientific politics, or for purely personal reasons." He wrote, "While I admit the thought is sometimes galling, that with all my life's work in hard places, its sufferings, dangers, and with the talents and ability, which others have seen, that I should be made to give up my scientific researches and forced into commercialism at my age of life, I shall pursue the tenor of my way, with the consolation of the knowledge that it is better to be wronged than to wrong." Just as it had been Cornell University's fault rather than his own that he had not completed the university's language requirements, in Hill's view it was outside forces, not his own decisions, that had forced him into the commercial world.

Puzzled and stung by the accusation of a personal grudge, Walcott replied that he held no grudge and had no personal ambitions, beyond work that he himself had begun years before Hill's involvement but set aside as he assumed administrative responsibilities. Of Hill's commercial work, he said, "The work has been unsatisfactory to you, and I will be frank to say, unsatisfactory to the Survey. I do not think that anyone can serve two masters and do it successfully. One must either devote his life to scientific work, with that solely in view, or, if he is to devote himself to money-making in connection with his scientific profession, his whole time must be given to that." Over time, the USGS would establish ethics rules expressly forbidding the kind of double life that Walcott decried. In Hill's day, however, there were no rules against it. Once again, even with a small mountain of later accounts, it is difficult know where the truth lay: Had unfair outside pressures forced Hill to take on commercial work? Or had it been his decision to take on commercial work, including writing articles for which he received compensation, to the inevitable detriment of his official USGS duties? Perhaps in this case the truth was somewhere in between. Even by the account of some close friends, during Hill's later years he struggled increasingly to pull together comprehensive reports without substantial editorial assistance.

Robert Hill on a field trip in Texas, 1905. (DeGolyer Library, used with permission)

The years between 1903 and 1912 marked an especially bitter time in Hill's life, which he seldom mentioned in later personal writings. Still, he made his way forward with the same dogged determination that had carried him from Tennessee to Texas to Cornell. Over the years following 1903, Hill's time and energies remained focused on commercial ventures of various sorts, mainly as a consultant on mining operations in Mexico. Almost all of his scientific publications during these years were published in mining journals, focusing on mining-related issues in Mexico. At least for a time, the work brought Hill financial reward if not great personal satisfaction. A 1904 article in the *Corpus Christi Caller Times* described Hill as "one of the most eminent scientists in the world," a successful businessman with "offices in New York, London, and Paris." The article noted that business interests would soon bring Hill to Mexico, giving him the opportunity to visit Texas. "As is well known," the article continued, "Hill is a Texan, and as loyal to the old state as he ever was." Of Hill's ventures into commercial geology, the article noted, "He is said to have made more out of one fee last month than he would have made out of two years' salary as a government official."

As Hill pursued business ventures, USGS officials and Alexander Agassiz at Harvard continued to look to him to complete unfinished reports on, respectively, the Trans-Pecos region and the Windward Islands. Hill himself was reportedly eager to complete the Windward Islands report; in 1906 he planned to return to the project and wrap up the report following an upcoming trip to Mexico. It was not to be. Hill carried some of the chapters with him on his Mexico trip, and somehow the manuscript was lost. "The inexplicable Jinx that has at times upset my life must have taken a hand," Hill wrote. The saga continued to play out, with a protracted back-and-forth with Agassiz over the manuscript Hill finally produced, which was not in adequate shape for publication. "I do not propose to edit any publications of the Museum," Agassiz wrote to Hill in May 1907. By late 1907, Hill's formerly cordial relationship with Agassiz ended—yet another relationship that ended on an unhappy note.

Bitterness also tinged much of the decade that Hill spent in the business world. Although his fortunes seemed to soar for a time, he found himself increasingly embroiled in battles with business partners and bookkeepers alike. His personal life offered no respite from the turmoil. During this time, still married but eventually entirely estranged from his first wife, Hill met and fell in love with another woman, Margaret McDermott. On August 15, 1911, their daughter, Jean, was born. Three months later, a business venture in which Hill had invested much of his wealth, the Camp Alunite gold mine, went bust when the mine proved to have no appreciable reserves of gold.

Having scaled great professional heights only to reach a nadir of his life at the age of sixty-three, Hill turned back to religion. While in (inevitable) fashion he railed against the dogma of the organized church, Hill's religion focused on the relationship between man and the universe, of which a strong faith in a "great First Cause (God, if you please)" was fundamental. Armed with newfound spiritual convictions and a new daughter, Hill left New York and moved with his second family to Los Angeles, where Jean could be raised away from the taint of illegitimacy. By all accounts, whatever failings Hill might have had as a man, he excelled as a father; from the start and through thick and thin, his two daughters were the unwavering bright lights of his life.

Hill was not the first man to choose California for a fresh start in life. His choice, however, had likely not been based only on a starry-eyed ideal of a land of new beginnings. Oil was discovered in the Los Angeles area during the closing years of the nineteenth century. During 1910, total oil production in the greater Los Angeles region reached 78 million barrels and was poised to go much higher. Hill's cadre of friends in Texas included industry geologists, some of whom had made their way west as the Southern California oil boom got underway. In moving to Los Angeles, Hill would not be without friends; moreover, he would be in a place where a talented field geologist could find professional opportunities.

Around the same time that Hill's tenure with the Survey came to an end, at least temporarily, on an unhappy note, in 1903 his old nemesis, Bailey Willis, faced frustrations of his own. If Hill had been unhappy to work under Willis's direction, Willis had been no happier with his increasingly burdensome administrative responsibilities, which left him little time for science. In 1903 a doctor diagnosed him with consumption, now known as tuberculosis. He requested a field assignment in the hopes of improving his health. In 1903 and 1904 Willis took leave from his administrative responsibilities to lead an expedition to China, sponsored by the Carnegie Institution. Following in his old mentor Raphael Pumpelly's footsteps, Willis and his assistants traveled down the Yangtze River, investigating the economic and structural geology of the region.

Following his adventures in China, Willis returned to the Survey, freed up from administrative responsibilities and more able to focus on fieldwork and writing. In 1910 another opportunity beckoned, and Willis embarked on his next grand adventure, again taking leave from the Survey to serve as a consulting expert to the Argentine government, which at the time was interested in exploring the mineral resources and irrigation potential of northern Patagonia. Whereas Robert Hill had crested his sixth decade of life in a sea of despondency, for Bailey Willis these same years were among his glory days. When he penned his memoirs, which he termed "a bit of autobiography," decades later, he titled the book *A Yanqui in Patagonia*. To the years before 1910, Willis devoted thirty-eight pages;

the years after 1914 received even less attention. The bulk of the book focused on his adventures in South America.

Willis's work in South America came to an end at the beginning of 1915, when the president of Argentina refused to approve the sizable contract to support Willis's work. "The bottom," Willis wrote, "had fallen out." Later that year, at the age of sixty-seven, Willis formally retired from the USGS, with whom his association had been remote in recent years, and accepted an appointment as professor and chairman of the geology department at Stanford University. And thus was Willis's professional orbit put on another collision course with Hill's.

In another parallel with Hill, Willis moved to California with his second wife, a woman who even shared the same given name as Hill's partner. But in this as in so many other respects, the two men's personal lives were a study in contrasts. Whereas Hill's Margaret was not legally his wife when they moved to California, Willis had married Margaret Baker, a talented artist and the daughter of the superintendent of the National Zoo in Washington, DC, in proper fashion, two years after the death of his first wife, who had struggled for years with health problems. The Bakers had been longtime friends of the Willises; Margaret sent her condolences to Bailey following the death of Altona. A correspondence, and romance, blossomed. The salutation of Willis's letters progressed quickly, from "Miss Baker" to " Marjorie" to "Dear little wife (to be)." Willis signed the "dear little wife (to be)" letter "Lovingly, Your Old Fogy." When the two wed in 1899, Margaret was twenty-four to Willis's forty-one. Willis's surviving daughter, Hope, had lived with Willis's sister Edith following her mother's death and then with her father after his remarriage, after which he and Margaret had three children. The eldest, their son Cornelius, arrived nine months and five days after the wedding; a second son, Robin, followed twenty-one months later. Their youngest child, a daughter also named Margaret, was born in 1908. In 1915 Willis moved with his family onto the Stanford campus, where they remained for many years—and where one particularly prickly ghost of Willis's early career would soon come back to greet him.

Chapter 4

CHAPTER 5

GOLDEN STATE

The chief line of study, in my opinion, is to make a
complete geological map of Southern California with
special attention to the fault lines, past and present.
—*Robert T. Hill*

By 1915, the Golden State had beckoned to both Robert T. Hill and
Bailey Willis for different reasons. For Willis, California offered a
plum to top off a successful scientific career: the opportunity to work as
a professor and pursue lucrative consulting opportunities, as well as to
turn his attentions to the geology of the West. For Hill, on a personal level
the West offered a combination of professional opportunity and escape. At
least initially, however, the man might have been taken out of Texas, but
Texas had not been taken out of the man: Hill spent long days in Los Angeles working on the Trans-Pecos report, as well as doing private consulting
work. At the end of 1912, he noted in his diary that he had completed "the
29th and final draft" of the report. After thirteen years of delay, the
Trans-Pecos monograph was delivered for publication by the US Geological Survey. Although the report would never be finished and published, the

Survey drew from Hill's draft to prepare its geologic map of Texas, and the delivery of the monograph may have paved the way for some degree of rapprochement with the agency. In June 1914, the USGS contracted with Hill to map the geology over 8,000 square miles (almost 13,000 sq km) in Southern California. For a time, while Hill focused on mapping in the Southland, Bailey Willis spent time in the field mapping parts of Central California, including the Lompoc area along the coast.

Their personal situations and motivations for moving to, and working in, California differed sharply; for its part, California beckoned to the two accomplished and by then senior geologists for different reasons. Just as the nascent USGS and its mandate to map the geology of the United States had inevitably drawn in talented and energetic young geologists thirty years earlier, by the early decades of the twentieth century, California beckoned to experienced geologists with different sorts of opportunities.

In the San Francisco Bay Area, the reality of earthquake hazard had been brought home to settlers even before the gold rush, as it became clear from the start that earthquakes were especially common in the Golden State. Fragmentary records describe a severe earthquake in June 1838; recent investigations have found the geological signature of this earthquake along the San Andreas Fault near San Juan Bautista. There is far better documentation of the earthquake that struck on October 21, 1868, now estimated to have been magnitude 6.5–6.8. This temblor caused heavy damage in the East Bay Area, leading some to refer to it at the time as the great San Francisco earthquake. Thirty-eight years later, the 1868 earthquake relinquished its title when *the* great San Francisco earthquake struck on April 18, 1906. The Richter scale, which provided the basis for all later magnitude scales, is famously logarithmic, a mathematical device Richter used to collapse the enormous range of earthquake sizes down to a comfortable range of numbers. With a magnitude now estimated at 7.8, the 1906 earthquake released about thirty times more energy, give or take, than the 1868 earthquake, with a fault break stretching roughly ten times as far.

While the 1906 earthquake focused societal as well as scientific interests squarely on the earthquake problem, in the San Francisco Bay Area as well as elsewhere in the world, earthquake science existed before 1906. By this time, the University of California, Berkeley had established itself as a center of earthquake studies, having operated early seismometers from the late 1880s onward. In 1890 the university hired a young Canadian geologist, Andrew Lawson, as assistant professor of mineralogy and geology. A towering figure in the field, Lawson was another scientist who by all accounts was not hurting for self-confidence.

Across the bay, Stanford University had also recruited scientists who worked on the earthquake problem. In 1891, Stanford president David Starr Jordan chose Hill's longtime colleague and confidant, John Branner, to be the first professor of geology at the university. When the 1906 earthquake struck fifteen years later, Branner found himself at the epicenter of a maelstrom. He investigated the fault line that had torn almost literally through his backyard, and with Lawson went on to establish a commission of scientists to investigate the massive earthquake. In the years that followed, Branner devoted considerable energy to the Seismological Society of America, created in the immediate aftermath of the great earthquake with the mission of understanding earthquakes and promoting risk mitigation. Elected president of the society in 1910, Branner was a driving force behind its new journal, the *Bulletin of the Seismological Society of America*. Although he declined to be reelected as president in 1912, he maintained effective control of the society and its journal for many more years.

Tapped by Jordan to take the reins as Stanford University president in 1915, Branner in turn recruited Willis. Where Hill had seen Willis as a contemptible Yankee who "hated all things southern," Branner's professional and personal relationship with Willis appears to have been cordial; the younger man's earlier views of the elder as "slow at ideas" apparently either evolved or was kept diplomatically under wraps. During his years at Stanford, Willis would, however, clash on many occasions with Andrew Lawson, their spirited scientific debates becoming the stuff of lore and legend.

At the age of sixty-seven, Willis might have settled into comfortable waters during his golden years, taking opportunities to quietly earn a good living as a consultant and work on interesting geological problems. But even in his later years, it was not in Bailey Willis's nature to sit still. He charged into the next chapter of his life, as keen as ever to explore new horizons. Working alongside Branner, he was drawn naturally into not only earthquake issues but also the Seismological Society of America. His professional interests expanded from a focus on geological structure to the short-lived geological paroxysms that, he and his colleagues by then realized, could so profoundly shape the landscape. Although he had never studied earthquakes in any serious way, as a longtime champion of Progressive causes he gravitated naturally to the cause of risk reduction.

To the south, in Hill's adopted hometown of Los Angeles, different geological challenges and opportunities presented themselves. Today Los Angeles is known for many things, some more positive than others. A diverse range of businesses—from health care, information technology, and aerospace to entertainment to massive service and hospitality sectors—now fuels the economy. Present-day Angelenos might not realize the extent to which Los Angeles was, in its early days, an oil town, its growth fueled by the initial discovery of oil near downtown Los Angeles in the closing years of the nineteenth century. Larger discoveries east of downtown, in what is now the Whittier area, followed the more modest initial oil strikes closer

Oil derricks adjacent to Huntington Beach, California (undated, early 1920s). (Orange County Archives, used with permission)

to the city. Oil production ramped up more sharply after a string of productive oil fields running across the western Los Angeles basin, following what was initially known as the Newport-Inglewood structural trend, was first discovered in 1921, with subsequent fields discovered in quick succession thereafter. By 1930, the greater Los Angeles region produced nearly a quarter of the world's oil, and the population crested one million. Oil derricks sprouted like dandelions in the California sunshine, along with imported palm trees, defining the Southland cityscape. In Huntington Beach, swimmers enjoyed a day at the ocean while an army of derricks churned and groaned immediately inland of the beach.

To this day, geology can be a boom-and-bust profession because the oil and gas industry, which employs so many geologists, is a boom-and-bust industry. The opening decades of the twentieth century were a good time to be a geologist. Hill was scarcely the first geologist to map the large-scale physiographic provinces in California but, as had been the case earlier in Texas, much basic work needed to be done to produce broad-scale maps. In this work, as before, Hill was not working as a petroleum geologist per se; rather, his work established the basic big-picture frame-work, including the locations of major faults, that could guide petroleum geologists' detailed investigations. Hill and other USGS geologists also undertook more focused investigations of known oil fields, work that paved the way for commercial exploitation. Hill's role, working under contract with the USGS, suited him well in at least some respects, leaving him to work on his own without day-to-day managerial interference—although that didn't stop him from complaining at times about the solitary nature of the job. As he had done throughout his life, Hill grumbled but persevered. Within the first year, he more than doubled the size of the initially defined study area, having realized that he needed to understand the adjoining regions to understand the geologic boundaries within his central study region.

Hill's professional life might have been satisfactory, more or less, but he could not escape the turmoil that had marked his life since early childhood. His wife, Jennie, died in November 1913, leaving him free to marry Margaret McDermott. Within weeks of the marriage, however, the relationship deteriorated. Finding herself pregnant again, Margaret lashed out at her new husband, at times leaving him, at times staying and brooding in silence. Vowing she would bear him no more children, she took drugs with the intent (apparently successful) of inducing abortion. Eventually Hill came to believe that Margaret was suffering from mental illness. He still loved her, but he could not find a way to quiet her demons nor make the relationship work. He remained, as he would throughout his life, a loving, supportive father, both to his first daughter, Justina, by then studying biology at Smith College, and to young Jean, his daughter with Margaret. The turmoil took a toll, however, on his work as well as his spirits, and he again faced financial struggles. His contract with the USGS, which had provided his main source of income for several years, ended in 1917 with the completion of his work in Southern California and an initial draft of a comprehensive report summarizing his findings. The draft was encyclopedic, stretching to 2,000 typed pages and a set of maps and plates that described and classified all of the major geologic features in Southern California.

Following the completion of the initial draft of his report, with responsibilities to support both his daughters, Hill found opportunities in the oil industry. With oil booms underway back in Texas and then in California, Hill found gainful employment as a consulting geologist for the oil industry. He set up an office in Dallas and once again spent time shuttling between states. In addition to his work as a consultant, he invested money in a number of ventures. While he never struck it rich with any of these investments and lamented working as a consultant rather than focusing on the basic science he longed to pursue, he was at least comfortable—and gratified to see the fruits of his labor being put to use toward productive ends. His professional reputation was also by that time well established.

Financial stability did not, however, bring domestic harmony. Although he still loved Margaret and remained as devoted as ever to their young daughter, Jean, the relationship continued to deteriorate. At the end of 1920, she left, taking Jean and all of their savings with her. She returned the next year, but she left again, permanently, in 1922, leaving Hill distraught over not only her loss, and not only fearing for the welfare of his young daughter, but also concerned that Jean might inherit her mother's mental illness. This concern apparently led him to reach out to his brother Jesse to explain the nature of Margaret's condition and seek a better understanding of relatives Hill himself had barely known. In a letter to Jesse, Hill acknowledged his own "irascibility," ascribing it to a "Wilkins curse." "Psychologically," he wrote, "it is a tendency to irritability following fatigue reactions to overwork." Hill went on to describe this tendency as hereditary and congenital, which "should not be treated as a moral delinquency." He also reflected on how circumstances had conspired with genetic predisposition, almost from the start of his life. Yes, he had accepted waking up at 2:30 in the morning for the paper route he took on at age nine, but at what cost? On another scrap of paper preserved among his papers, he typed out a quote by Adela Rogers St. John: "I wonder if the children of a motherless childhood are ever quite the same as other people. I wonder if the scars of small hungry souls feeling blindly for love ever quite heal."

As a poignant aside, Robert Hill lived long enough to see the younger of his two daughters, Jean, educated at the University of California, Los Angeles and married to classmate Harold Guttormsen. Two years after the marriage, Jean died in childbirth when her own daughter, Carolyn Jean, was born. For a time before Guttormsen remarried, Carolyn was raised by Jean's mother, Margaret—the woman in whose care Hill had feared for Jean's own welfare. The arrangement did not last long: when Guttormsen remarried the following year, he and his new wife reclaimed his young daughter. Especially for a baby born in 1858, Robert Hill lived a respectably long life, but he did not live quite long enough to know that his only grandchild became a motherless child on the day she was born.

As he came to appreciate the extent of his wife's condition, Hill could not have imagined the cruel twist that fate had in store for his younger daughter, but he could imagine other perils in her future. In reply to Hill's letter, Jesse responded with a thoughtful letter of his own. He expressed sympathy about Margaret's condition, suggesting that perhaps her sister might be able to reason with her during her "sane periods." In retrospect,

Hill's description of wildly mercurial moods punctuated by periods of normalcy suggests that she might have suffered from bipolar disorder. There is no indication that Margaret ever sought medical treatment, however, and even if she had, treatment options for mental illnesses remained limited at best in the early twentieth century. To the suggestion that their mother's condition had been hereditary, however, Jesse begged to differ. Describing several relatives on their mother's side of the family as gentle, even-tempered pillars of the community, Jesse wrote, "There was never a case of insanity prior to our mother's in the family." Her breakdown, in Jesse's view, had been a consequence of grotesquely inhumane circumstance. "Do you realize," he wrote, "that Tom was barely fifteen when he enlisted as a soldier, and Sam was but thirteen. That I ran away three times in an endeavor to join the army—on one of which trips I was absent about three weeks. It was such things as these, coupled with dire poverty due to confiscation of all our father's resources . . . that caused our mother's affliction." Jesse's own understanding of their mother's breakdown came in large part from confidences shared by one of their mother's sisters, Maggie, who came to live with the family the same year Robert had been born.

Jesse went on to discount any suggestion of a "Wilkins curse." In his view, the predominant family trait on that side was pride: "Not a false pride, but a pride such as is characteristic of the best type of an English gentleman." Jesse went on to say, "It is because of this . . . that you were inspired to achieve things—not money." The traits they had inherited from their mother's side had, in Jesse's estimation, "made Tom a dreamer, Joe an editor, and you a geologist." Of their father, Jesse offered different views: "Our father," he went on to write, "was himself 'high strung' and quick to take offense." Jesse further responded to Hill's musing on where his own "gambling spirit" might have come from; in Jesse's view, he did "not believe that the gambling spirit . . . was inherited."

That Hill himself was irascible and inclined toward a sense of persecution, by whatever combination of nature and nurture, he did acknowledge by the later years of his life. There was also no question, however, that nurture had conspired in unfortunate ways to stoke whatever tendencies nature had bestowed on him. Yet through it all, from the age of nine and even before, there remained also other traits, including a keen intellect, an innate ability to understand the landscape, a passion for teaching, and the determination of a (compact) freight train; these traits had propelled him from Tennessee to Texas, from Texas to Cornell, from Cornell to Washington, DC. Eventually, they brought him all the way to California.

In California, Hill's trajectory also collided, perhaps inevitably, with the earthquake problem. An old-school geologist to an even greater extent than Willis, Hill also had little interest in earthquakes, which he viewed as fleeting disturbances. While his experiences on Mount Pelée had sharpened his appreciation for the potential dangers of geological hazards, in his experience the impacts of earthquakes were far more modest than the toll taken by other types of disasters. This perspective shone through his remarks about earthquakes, when he was asked to comment on them. As early as 1909, Hill had been called upon to weigh in on the subject after an earthquake of an estimated 7.1 magnitude struck the Strait of Messina, at the toe of Italy's boot, leaving tens of thousands dead. The disaster had quickly given rise to alarmism, with dramatic headlines again accompanied by grim photographs, the media coverage focusing on the most spectacular instances of damage. Moreover, then as now, earthquakes are not preceded by predictions but, rather, the reverse. Which is to say, anytime a damaging earthquake captures the headlines, talk, if not actual predictions, of possible future calamities—up to and including end-times—inevitably surface.

Quoted extensively in the *Arkansas Gazette*, Hill explained, "Earthquakes, as terrible as they may seem when they destroy a few thousand human beings, are but one of the many manifestations which we have of the world at work." The first scientist to have witnessed a Pelean eruption continued, "These processes are not destructive by any means; they are the growing faculties that, instead of imperiling the life of the earth, establish its indestructibility." When Hill spoke of destruction here, he did not mean it in the sense that most people would—that is, destruction to life, limb, and property. That earthquakes could pose a real danger and take a serious toll he always acknowledged, as evidenced by his unpopular support for a canal through Panama rather than the more disaster-prone Nicaragua. When Hill commented in this article, he spoke with a geologist's global perspective and in response to alarmist talk that end-times were nigh, his words a statement that neither earthquakes nor volcanoes could ever destroy the planet. He also spoke as a man who had seen the tolls taken by the 1900 Galveston storm and 1902 eruption of Pelée.

In the eyes of a geologist concerned with the grand tapestries of the planet, earthquakes were dangerous, yes, but not of enormous consequence, and the risks could be mitigated. He again made his perspective clear in the opening paragraph of an article that he published, presenting some of the observations from his earlier fieldwork, in the *Bulletin of the Seismological Society of America* in 1920 titled "The Rifts of Southern California." In this article Hill wrote, "Earthquakes are the least harmful of any of the various and usually terrifying manifestations of nature—much less so than the disastrous consequences of lightning, hail, flood, and tornado, each of which probably annually kills more people in the United States than earthquakes have killed in the entire history of Southern

Fence north of San Francisco offset laterally by the 1906 San Francisco earthquake. (USGS photograph)

and the faults themselves are largely sealed and closed." He concluded on a nearly rhapsodic note, "In conclusion I must repeat that these great rifts, traceable for hundreds of miles, are to me the most interesting and wonderful physiographic phenomena I have ever had the pleasure of seeing." And yet he went on to dismiss the importance of the earthquakes that had carved out these spectacular features. "All I can say," he wrote, "is that [tremors] sometimes occur. San Jacinto, Brawley, Lytle Creek, Elsinore, and other places have each had their little tremblings, but the activities of life and property go on."

Hill's inclination to downplay the severity of earthquake hazard in California was based in part on his conclusion that overall fault motion in California was much lower than it had been in earlier geological times.

In Hill's view, the "most interesting and wonderful" rifts had been created by forces of the distant past. Geologists now know that the San Andreas Fault has been cooking along at about the same rate for a good 12 million years, give or take, if not more, with no signs of recent slowing. Hill's conclusion, however, was not unreasonable at the time, given the mistaken paradigm that virtually all geologists accepted. That is, if one focuses only on vertical faulting, geological processes in California would certainly appear to have slowed in recent geological times. Before about 30 million years before present, the configuration of the plate boundary was different; the primarily compressional large-scale forces created California's spectacular mountains. The rate of mountain-building in California has indeed slowed dramatically over time, starting from the time the lateral San Andreas Fault system began about 30 million years ago. In this and other respects, looking back at things written by Hill and his contemporaries, one finds that some glaringly wrong statements were understandably wrong, and sometimes even insightful, at the time.

In any case, Hill's article ended on a somewhat different note. "In conclusion," he wrote finally, "I cannot help repeating the deduction of Dr. Branner, that the best way to master the fear of earthquakes is to study them and to familiarize ourselves with them. The chief line of study, in my opinion, is to make a complete geological map of Southern California, with special attention to the fault lines, past and present." These concluding words sound discordant, coming at the end of an article that had described many of the state's faults but described earthquake hazard as inconsequential. In Hill's view, however, a better scientific understanding offered the best hope of overcoming what he saw as irrational fear of earthquakes. As he said in a speech to the Los Angeles Rotary Club following the locally damaging 1920 Inglewood earthquake, "Fear is the cruelest of all devil (or evil) made human obsessions, and this emotion itself has a death list compared to which that of the earthquake is utterly insignificant."

Moreover, while Hill discounted the dangers posed by California faults, as an industry geologist he knew that mapping the state's many faults was of enormous practical importance. By bringing different rock units in juxtaposition, faults not only create topography, they can also create so-called seals across which fluids are not able to flow easily. Thus in many areas, including California as well as parts of Oklahoma and Texas, faults create many of the subsurface traps in which oil and gas are found. Like Bailey Willis, Hill had always been motivated by two things: a passion for science and an abiding conviction that geology could contribute to human welfare. Whereas Willis gravitated toward so-called Progressive causes, Robert Hill, a man who had known real poverty and hardship during his formative years, focused on potential contributions of his science to economic welfare. Thus did Hill and Willis take different paths to arrive at the same place; thus did Hill's article, which started with a dismissal of the importance of earthquakes in California, conclude with the suggestion that it was of utmost importance to compile a map of California's geologic faults.

CHAPTER 6

FRAMING THE DEBATE

> The happy impute all their success
> to prudence and merit.
> —*Jonathan Swift*

As Robert Hill continued to find gainful employment as a consultant to the oil industry in California as well as back in Texas, John Branner and others continued to attack, and call attention to, the earthquake problem in California. From the start, another scientist played an important role in this part of the story: Harry Oscar Wood. Wood—whose tight-lipped, steady-eyed photographic portraits all bring to mind a single word, "earnest"—arrived on the scene via a different trajectory than that of Hill and Willis, as well as their more senior colleagues, including Branner. Born in Maine in 1879, Wood earned his bachelor's and master's (although not a PhD) degrees from Harvard before accepting a teaching position at the University of California, Berkeley in 1904. Like Hill, Wood's degrees were in geology. Seismology effectively didn't exist as a separate field of study, at least in the United States, in the late 1800s. Early seismology programs, including those at Stanford and UC Berkeley, were established by geologists. Moving to an earthquake-prone part of the country and soon witnessing one of the most momentous earthquakes to ever hit the United States, Wood himself became part of a transformation. After the 1906 earthquake struck, he investigated damage patterns in San Francisco; in the years following the earthquake, he taught the first university course in seismology in the United States.

In 1908, pioneering geologist G. K. Gilbert—a visionary scientist and one of Willis's mentors and heroes during his early days at the US Geological Survey—tapped Wood to compile a map of potentially active faults in Northern California. Later at UC Berkeley, Wood was tasked with the job of overseeing the university's seismometers, at that time behemoth instruments tuned to record waves generated by big earthquakes around the globe—waves that were imperceptible to humans. In spite of his contributions in teaching and research, Wood's career advancement in academia was hampered by his lack of a PhD. He remained a low-paid instructor, with no hope of promotion. He left Berkeley in 1912 for Hawaii, where he worked at the newly founded volcano observatory on the Big Island. During his years in Hawaii, he developed a relationship with Arthur Day, director of the Carnegie Institution's geophysical laboratory, who conducted research in Hawaii during Wood's tenure there.

Wood had been born comfortably after the Civil War dust had settled, but another war intervened in his life: in 1917 he joined the Army Engineer Reserve Corps and moved to Washington, DC, working at the National Bureau of Standards, for the duration of World War I. During this time, he had the opportunity to interact with influential senior colleagues, including paleontologist John Merriam. At the Bureau of Standards, Wood worked on the development of a piezoelectric seismometer that was being tested to locate cannon fire. Because seismometers record ground vibrations, and more things than earthquakes cause the ground to shake, it was neither the first nor the last time that strategic national interests spurred development of modern instrumental seismology. After the war, in 1921 Merriam left his role as dean of faculty at Berkeley to become president of the Carnegie Institution.

On a tangential note, one doesn't get too far into the annals of early earthquake exploration in the United States before realizing just how small the community of earthquake scientists was. As noted, the 1906 San Francisco earthquake did not mark the start of modern seismology, as some like to claim. The fact that the earthquake had been recorded by seismometers around the globe tells us that the field of seismology existed before the earthquake. It would be fairer to say that, in the United States especially, the field of seismology came of age with the 1906 earthquake.

Harry Oscar Wood (undated). (California Institute of Technology Seismological Laboratory, used with permission)

With essentially no focused seismology programs, or seismologists, in the country before the earthquake, it fell to a small group of individuals to launch research and other programs, including local earthquake monitoring. Earlier chapters focus on two of these individuals and how their orbits intersected from the start of their careers.

In the late 1800s, the field of geology as a whole was not terribly big in the United States. The development of seismology in the United States was propelled by the efforts of a small group of individuals whose careers intersected at a handful of key institutions, including Stanford, the University of California, and the Carnegie Institution. These individuals hailed from different generations: an old guard, including John Branner as well as John Wesley Powell and Clarence King; the generation of geologists, including Hill and Willis, who were starting their careers when the USGS was founded; and the younger generation, including Harry Wood, that followed. The orbits of these men—and they were virtually all men—crossed at different times and in different places. For better and for worse, their interpersonal relationships shaped not only individual career trajectories but also the development of the field. John Merriam's trajectory would prove to be especially important for seismology: taking the helm of the Carnegie Institution in 1921, he had considerable say over funding decisions. Through what would prove to be a pivotal period for earthquake science in California, Arthur Day was another significant figure. Initially employed by the USGS in 1900, in 1907 Day became director of Carnegie's Geophysical Laboratory, a position he held until his retirement in 1936. In this capacity he corresponded regularly with Wood, with whom he had a warm personal relationship. Their conversations ranged from technical matters such as instrumentation to political issues. Day also had a close professional and personal relationship with Bailey Willis.

In Harry Wood's case, his career would be influenced in critical ways by a number of the relationships he developed with a few colleagues, including Day and Merriam. During his years in Hawaii and then in Washington, DC, Wood never lost interest in the earthquake problem back in California. Years after Wood's death in 1958, Wood's longtime colleague Charles Richter said, "Seismology owes a largely unacknowledged debt to the persistent efforts of Harry O. Wood for bringing about the seismological problem in Southern California." While still on the East Coast, Wood published a pair of papers in 1916, one titled "California Earthquakes: A Synthetic Study of Recorded Shocks" and the other "The Earthquake Problem in the Western United States." The first paper included a map of fault lines in California, drawing from his own work and the work of earlier geologists.

Wood's second 1916 paper laid out the case for the importance of earthquakes in California. Whereas Hill would later downplay the importance of earthquakes in his own article in the same journal, Wood laid out a diametrically opposed case. He noted that much of California had been rocked by great earthquakes in the past. "There is overwhelming geologic evidence," he wrote, "that here great earthquakes have originated again and again in localities which overlap each other along recognizable belts—'living' zones of geological faulting." Unlike Hill, Wood believed that California's spectacular geological structures went hand in hand with present-day earthquakes and earthquake hazard. Wood went on to emphasize, although not overplay, the latter: "Therefore it must be clearly recognized that throughout this large seismic province the occurrence of strong earthquakes is *in moderate degree* a menace to life and property—in a greater degree than many residents of the region are willing to admit. It is idle to deny this, or to strive to conceal it." Wood struck a generally prudent tone: "In justice to the future development of this great, important region," he added, "care must be exercised not to over-emphasize this danger, which, *as already stated*, is not great, nor, in a given locality, always threatening." Wood went on to describe efforts by earlier scientists to call attention to the seriousness of earthquake hazard. In the end, Wood's article concluded on a note similar to the one on which Hill's article ended, with recommendations for work that was urgently needed for California to come to grips with its earthquake problem, thereby beginning to mitigate earthquake risk. Chief among these recommendations was the development of instruments that could record not the imperceptible waves generated by distant global earthquakes but, rather, weak and strong shaking from local earthquakes.

Although there had been forerunners, if one looks to identify the opening salvo of the great quake debate, arguably it was the publication of Wood's two papers in 1916. It might have been viewed as the moral equivalent of Fort Sumter, except that the other side did not engage immediately in battle. These papers painted earthquake hazard as a serious concern not only for Northern California, but for the entire state. In the years following publication of these articles, and after the first world war, Wood teamed up with a few colleagues to develop the outline of his ideas into a full-fledged proposal to the Carnegie Institution. The proposal focused on Southern California, where the earthquake problem had been particularly slow to gain traction, in spite of the growing conviction of Wood and most of his colleagues that the entire state faced significant earthquake hazard.

When Wood wrote about denial and concealment of hazard, he did not choose his words idly. Fueled by the initial discovery of oil, Los Angeles had been on its way to its future position as the state's major economic hub when the 1906 earthquake grabbed headlines around the world. In the region hardest hit by that temblor, the "denial of disaster" ran up against stiff headwinds, including the news articles that had splashed across front pages around the world, accompanied by grim photographs. To the south of San Francisco, earthquake hazard itself could be more plausibly denied or, at a minimum, debated. Earthquakes were certainly not unknown in the Southland. Having moved to Los Angeles with his family in 1909, Charles Richter felt his first earthquake, a moderate shock near Riverside, in 1910. For most Southland residents in the years before and immediately after 1906, however, earthquakes presented at most a mild and occasional interruption of day-to-day life. The immediate Los Angeles area moreover remained quiet, for a time. From the period when Southern California's population first started to grow rapidly in the closing years of the nineteenth century through 1915, there is no evidence of any earthquakes larger than magnitude 4 within the central Los Angeles basin. By 1900, an earthquake in the Los Angeles area as large as magnitude 5 would have been recorded by early seismometers operated by UC Berkeley; no such event was recorded through the 1900s and 1910s.

After 1915, however, earthquake activity in proximity to Los Angeles picked up. A damaging earthquake struck Southern California on the afternoon of April 21, 1918, centered well east of the Los Angeles area in a then-remote region along the San Jacinto Fault. Now estimated to have been about the same magnitude as the 1994 Northridge earthquake (6.7), this quake caused serious damage to the mostly brick buildings in the business districts of San Jacinto and nearby Hemet. Ralph Arnold, a former student of Branner's at Stanford and by 1918 a wealthy petroleum geologist, investigated the earthquake and published a report describing the San Jacinto and neighboring faults. Waves from the sizable temblor reverberated throughout Southern California.

Ralph Arnold is another of the small handful of scientists who will make recurring cameo appearances throughout our story. In fact, perhaps no single individual played a bigger behind-the-scenes role in the debate as Arnold. To introduce him briefly, as a young man Arnold parlayed his geology degree from Stanford into great wealth as an oil man and thence parlayed wealth into political capital. Through the 1910s and 1920s he traveled widely across the United States, bouncing between his home in the Los Angeles area, New York, Washington, DC, and other parts of the country where he had oil ventures. His date books reveal that through the 1920s he met with not only business leaders in Los Angeles, but also then–Commerce Secretary Herbert Hoover and even President Calvin Coolidge himself. Arnold, like Hill, had been drawn to collecting fossils as a boy, but he grew up to be as politically savvy as Hill was impolitic. Arnold's small date books reveal somewhat more about the man than about his travel and appointments. The man of great wealth and political acumen was also an avid gardener, and sometimes he pressed small flowers and four-leaf clovers among the pages chronicling his meetings with some of the most powerful political figures of his day.

The 1918 San Jacinto earthquake, which helped focus Arnold's interests on earthquakes in Southern California, rocked the Southland in a literal and figurative sense. The earthquake was front-page news in the *Los Angeles Times*, with the front page of the April 23 issue sporting a banner headline, "TOTAL TEMBLOR LOSS THREE HUNDRED THOUSAND," accompanied by four front-page photographs of dramatic earthquake damage. Other articles on the front page attempted to balance the alarmist tone. One article, titled "TRUTH ABOUT EARTHQUAKE," noted that damage to Los Angeles had been light, amounting to only about a thousand dollars. "The great Los Angeles aqueduct and the irrigation system of the Imperial were not damaged," the same article noted. A second front-page story bore the headline "EAST ALARMED; IS REASSURED." This article presented the statement issued by Oscar C. Mueller, president of the Los Angeles Chamber of Commerce: "Many telegrams reaching the Chamber of Commerce and individuals from the East indicate that greatly exaggerated reports of damage from earthquake in Southern California are being circulated. We resent these false reports. Thorough investigation of the chamber here shows the damage in Hemet and San Jacinto not nearly so great as was first reported, and that the damage in Los Angeles consists of a few broken windows. All persons having relatives in this section," the statement concluded, "may rest assured of their safety."

Much about the great quake debate might seem patently clear on its face, including the notion that, from the start, scientists like John Branner and Bailey Willis were the heroes in white hats, crusading for improved societal resilience and public safety, while business leaders and city boosters like Mueller endeavored to downplay earthquake hazards in hopes of advancing business interests. It is an easy narrative to tell, an easy narrative to believe. The easy story is, however, not always the right story.

Business leaders like Mueller sometimes made valid points. In modern as well as earlier times, news reports of damage from large earthquakes zoom in with laser focus on the sites of especially spectacular damage. A building that does not fall down is not news, whereas a photograph of a striking instance of collapse garners great interest and attention. Disasters horrify us, but they also captivate us; we can't look away. This reality translates into a fundamental reporting bias that can give rise to greatly inflated perceptions of damage, a bias that has been noted and decried by thoughtful observers throughout time, within and outside of the community of earthquake professionals. For example, an article published following a moderately large Bay Area earthquake in 1865 noted, "One paper indulged in a tremendous display of exclamatory sensation headings: 'Tremendous Earthquake!' 'Probable Destruction to Life!' 'Immense Damage to Property!' . . . Any one visiting San Francisco on

Monday morning, would have expected to witness a scene of widespread desolation. Instead, however, to outward appearance everything seemed as usual. Portions of the walls of two or three miserably constructed buildings had fallen down, and many others were somewhat cracked and jarred." In recent times, a century and a half after 1865, headlines informed newspaper readers around the world that Nepal's cultural heritage had been destroyed by the 2015 magnitude-7.9 Gorkha earthquake. In fact, while the temblor did take a heavy toll on some of the country's remote villages, claiming nearly nine thousand lives, signs of earthquake damage were few and far between in many parts of Kathmandu Valley. Not all of the country's cultural heritage sites survived intact, but many did, as did the overwhelming majority of the valley's other structures. Getting back to Southern California, the 1918 San Jacinto earthquake itself was front-page news outside of California. The *New York Times* ran a front-page article headlined "EARTHQUAKE DOES HEAVY DAMAGE IN CALIFORNIA TOWNS," with subheaders continuing in the same vein, including "Los Angeles Crowds Stampeded by Two Shocks and Many Persons Are Hurt" and "San Bernardino Suffers."

What's more, now as in earlier times, outsiders' appreciation of California geography can be limited. Even modern-day Californians know what it's like to be on the receiving end of misplaced concern from out-of-state loved ones who hear frightening news from a part of the state that is actually hundreds of miles away from their own location. It can be easy for a person from the East Coast to not appreciate the sheer size of California, the fact that the distance between San Francisco and Los Angeles, for example, is not too much less than the distance between New York City and Raleigh, North Carolina. A headline mentioning crowd stampedes in Los Angeles following an earthquake more than 80 miles (about 130 km) to the east did nothing to improve readers' appreciation of

Damage in downtown Inglewood caused by the 1920 magnitude ~5.0 Inglewood earthquake. (California Historical Society Collection, reproduced by University of Southern California Library)

geography. In short, Oscar Mueller was not entirely wrong when he spoke of exaggerated reporting on damage from the 1918 earthquake. The public does tend to get an inflated sense of earthquake damage from media reports, and it was especially easy for an overly alarmist tone to shape people's ideas on the other side of the country. Even scientists can be guilty as well. When a scientist like Willis or Branner pointed to a recent illustration of earthquake damage or possibility of future hazard, there could be a fine line between calling attention to the problem and buying into, if not outright fueling, the hype.

By virtue of its modest impact and remote location, the 1918 San Jacinto earthquake was among the early battles in the great quake debate, but others soon followed. Over the following two years, earthquakes—including a January 1920 earthquake in the state of Veracruz, Mexico, that killed thousands of people—occasionally merited discussion in the *Los Angeles Times*. Within the city of Los Angeles and its immediate environs, earthquakes remained more of a minor curiosity, and perhaps nuisance, than a pressing concern. The sense of serenity was interrupted on the evening of June 21, 1920, when a moderate earthquake struck the city of Inglewood, southwest of the city of Los Angeles. This earthquake predated not only the advent of earthquake monitoring in Southern California, but also Richter's introduction of the magnitude scale in 1935. Still, coming on the heels of other quakes in the Southland, the size and location of the 1920 earthquake, and the tight concentration of significant damage it caused, drew the attention of the scientific community. At Stanford, John Branner asked his former student Stephen Taber, who had arrived at Stanford the day before the earthquake to spend his summer there as a visiting professor, to travel to Southern California to study the earthquake and local geology. Taber spent three days investigating the effects of the temblor as well as the lay of the land. He returned to the area again when a series of minor shocks in the Los Angeles area later that year caused slight damage.

To Taber's expert eye, it was evident that, while fault lines within Los Angeles city limits appeared to be short, they were clearly active. Noting subtle warping of the surface that aligned along a northwest-southeast-trending swath across the Los Angeles basin, he identified a zone that stretched some 85 miles (137 km) from the Santa Monica Mountains through and well beyond the southern corner of the basin. "Additional evidence," he wrote, "is needed before it can be determined whether this line, through its entire length, is due to faulting, but the probability of that hypothesis is strong." Los Angeles and its immediate environs sit atop a deep geological basin filled with young (on a geologic time scale) sediments, and those sediments showed clear signs of significant deformation, their young age notwithstanding. "The Inglewood-Newport-San

Onofre fault, or fault zone," Taber continued, first naming the feature now known as the Newport-Inglewood Fault, known earlier among petroleum geologists as the Newport-Inglewood structural trend, "is probably the longest [fault] in the southern California coastal region." Taber's conclusions echoed and sharpened concerns that others had expressed. Noting that "there is every reason for believing that earthquakes will be felt in the Los Angeles district as frequently in the future as in the past," he added the same refrain that Hill and Wood had ended their articles with: that it was "of utmost importance that all active faults . . . be located so that engineers and architects can take precautions in the location and construction of aqueducts, buildings, and other structures."

Modern investigations have concluded that the Newport-Inglewood Fault is not so young but, rather, has been active for millions of years. Yet as was observed from the early studies onward, the fault has no clear surface expression, as one finds easily for the San Andreas Fault. The complex nature of the fault zone is likely due to two factors. First, the geologically young sediments in the Los Angeles basin work against development of a clean fault at shallow depths. Imagine a fault zone at depth trying to break to the surface through a giant overlying sandbox. Secondly, geological investigations reveal that the Newport-Inglewood Fault might have been active for a long time, but never at a very fast clip. The long-term rate of motion along the fault, and therefore its long-term rate of earthquake production—is now estimated to be a small fraction, perhaps about one-fiftieth, of that of the San Andreas Fault.

Scientists do generally agree, however, with Taber's conclusion that the Newport-Inglewood is a major through-going fault and that this fault was responsible for the 1920 earthquake. And with the 1920 earthquake, though it was a moderate earthquake still believed to have occurred on the Newport-Inglewood Fault, the great quake debate gathered steam. To be sure, the seeds had taken root earlier. But the debate had mostly simmered in the background, tempered in part by World War I, which not only pulled Harry Wood (along with other scientists) to Washington, DC, but also focused the country's attentions on other problems. The 1918 San Jacinto earthquake, which had been close enough to stoke the flames in a minor way—and certainly fueled behind-the-scenes efforts by Branner, Wood, and others to focus attention on the earthquake problem—had not been consequential enough to hold the public's interest for long. Earthquakes remained mostly out of sight, and, willfully on the part of some, out of mind. Even before 1920, Branner had been frustrated that Angelenos were "afraid to say [the word] earthquake out loud"—as if saying the word might summon the demon from the underworld.

Branner embarked on a letter-writing campaign to drum up support for the Seismological Society of America and for earthquake monitoring, commenting to geologists, engineers, and businessmen on the "queer things" written in Los Angeles newspaper articles about earthquakes. Branner did not provide specific examples of the statements he considered "queer," but one does not have to look too hard to find articles that he might have had in mind. For example, an article published in the *Los Angeles Times* after the 1918 San Jacinto earthquake noted that "Earthquakes of the Southern California variety are much akin to ghosts; they occasion great mental anguish and vexation of the spirit, but their physical violence is negligible." The 1920 Inglewood earthquake itself, however, fueled the determination of city boosters: later that same year, the Los Angeles Chamber of Commerce created an earthquake committee, on which Ralph Arnold served along with a number of prominent industry geologists. The committee would remain active through the 1920s, eventually pulling Robert Hill into its ranks.

Chapter 6

One must note that, in this case as well, efforts to downplay earth-quake hazard in Southern California were never, of course, entirely suc-cessful. Then as now, earthquakes make for splashy news. Numerous articles appeared in newspapers describing the effects of damaging earth-quakes, including the 1920 Inglewood event; other articles called attention to past and/or likely future earthquakes. But, as was the case in the after-math of the 1906 and 1918 earthquakes in San Francisco and San Jacinto, respectively, the counterpoint appeared, early and often, as well. Just four days after the Inglewood quake, residents of the small town of Hemet, which sits in proximity to the San Andreas Fault east of San Bernardino, found two articles side by side in their local newspaper, the first headlined "Hard Temblor at Inglewood" and to its immediate right "No Danger Here of Earthquake." The latter article did not downplay earthquake hazard entirely, noting that "we had an adjustment in 1918." But overall the mes-sage supported the headline. "I have observed no great settlings along this portion of the San Andreas," the article quoted "San Bernardino earth-quake specialist" Orlis Kennedy as saying, "and none has been reported that would indicate local adjustments big enough to ruffle the mental equi-librium of the natives." Kennedy went on to opine, "There will be no great damage to Los Angeles property, but the population is high tensioned and might suffer mental shock from a slight mid-day tremor." (Perhaps this was another of the statements that John Branner had regarded as "queer.") Kennedy, an attorney by profession, had wide-ranging interests, including economic geology; he is credited with having originally conceived a project to drill for oil in the Painted Hills region west of Palm Springs.

By 1920 the battle lines in the great quake debate had mostly taken shape, with a small group of scientists including Wood and others pitted against others who, by all indications, sought—at times not unreasonably—to downplay the severity of local earthquake hazard. The latter group found some support from geologists and others like Kennedy with professional ties to the oil industry, but also others. At UC Berkeley, Andrew Lawson, among the leading geologists of his day, wrote in 1927 that "Los Angeles is situated a favored locality sufficiently distant from the active faults upon which great earthquakes are generated. . . . So far as destructive shocks are concerned Los Angeles has not acquired the habit, and if we may judge by the record, is not likely to acquire the habit." Most academic scientists, however, united behind a clear message: earthquake hazard posed a serious concern to Southern as well as Northern California, and earthquake monitoring and investigations were urgently needed to better understand and mitigate this risk.

The battle lines were never entirely precise, however, on either side. Just as Robert Hill's parents reportedly had some sympathy for Union causes, some industry and consulting geologists joined the call for improved monitoring and research. Among them was petroleum geologist Ralph Arnold, the former student of Branner who investigated the 1918 San Jacinto earthquake. Arnold initially became part of a chorus of voices calling for earthquake science to better understand hazard and mitigate risk, to the point that Harry Wood took exception to Arnold's "boosterism"—boosterism, in Wood's eyes, not for industry but for earthquake science. A consummate politician, Arnold endeavored, generally successfully, to get along with everyone. As battle lines deepen, however, sympathies fall by the wayside; every individual has a decision to make: Which side are you on? Which army do you join? Among the constellation of individuals involved with the great quake debate, Arnold's loyalties appear to have been uniquely ambiguous, although they perhaps emerged more clearly in time.

By 1921, fueled in part by what Wood referred to as "signs of reawakening seismic activity" in the region, the stars had aligned to launch earthquake exploration in Southern California. After John Merriam took over as president in 1921, the Carnegie Institution agreed to provide support for the program, with a budget of $25,000 and another $3,500 earmarked for an Advisory Committee in Seismology headed by Arthur Day. It fell to this prestigious committee, whose members included Ralph Arnold, Bailey Willis, Andrew Lawson, and California Institute of Technology's Robert Millikan, to guide the development of a Carnegie program. In 1921, Millikan himself created an advisory body "as a fact-finding committee for collecting data, initiating researches, formulating principles of design and construction, and furnishing information of value to engineers, architects, city administrations, chambers of commerce, and others interested." The membership roster of this committee included familiar names: Ralph Arnold, Arthur Day, Bailey Willis, and Andrew Lawson.

With Carnegie support, the Pasadena Seismological Laboratory was launched under Wood's direction. The lab operated initially in borrowed office space at Mount Wilson Observatory, in the San Gabriel Mountains overlooking Pasadena. Wood teamed up with astronomer John Anderson to develop a new kind of seismometer, a small and elegantly simple device designed to record local earthquakes. The instrument, first described in a 1925 article authored by Anderson and Wood, went on to be known as the Wood-Anderson torsion seismometer. Test instruments were installed in the early 1920s; the first permanent monitoring station outside of Pasadena would be installed in Riverside in 1926. During the early development of the monitoring network, John Branner leaned on Ralph Arnold to solicit financial contributions from industry. Although Arnold's fundraising was not always successful, he did make inquiries to help find sites where network stations could be established.

Also in 1921, the US Coast and Geodetic Survey received a $15,000 appropriation to undertake triangulation surveying of California, work aimed at detecting strain in the earth's crust that might point toward future earthquakes. The appropriation had been added by Herbert Hoover, in his then-capacity as commerce secretary under President Calvin Coolidge. Hoover not only had studied geology at Stanford University and had himself undertaken triangulation surveying as a classroom exercise, he also had married the only female geology major at the time, Lou Henry.

By the fall of 1923, as the Carnegie-supported program inched forward, Wood and his colleagues began to reach out to community leaders. In September 1923 they invited several members of the press to attend a series of talks they were to present to the Pasadena Chamber of Commerce about earthquakes and the Carnegie Institution's proposed program. Although Wood later wrote to Day that the talks seemed to be well received, he was not happy with the publicity that followed. In an article published in the *Los Angeles Times* on September 7, 1923, readers were told that "Dr. Arthur L. Day . . . feels that California should give more publicity to her earthquakes." The article went on to describe the exaggeration of California earthquakes by the outside press. It continued, "For our part . . . we have never regarded Southern California's little quivers as anything but an innocent expression of the joie de vivre, tremors of delight, the natural exuberance of gay and healthy youth." Wood passed along news clippings to Day, who replied, "If that is a fair sample of what the responsible newspaper men, handpicked for the purpose, can do in Los Angeles, then the less publicity we get the better."

Sometimes a historian is left to speculate about forces that might have played out behind the scenes. Sometimes, no speculation is needed. On October 8, 1923, not too many weeks after the Pasadena Chamber of Commerce presentations, Wood wrote again to Day to relay his recent experiences at a breakfast meeting to which Dr. Hubble, of later Hubble telescope fame, had "carried [him] off." At the meeting, Wood found a gathering of "some interesting and influential people in this region. . . . Two of these men," he wrote, "proved to be very substantial and solid citizens—long-time residents of California—one of whom impressed me more than any public man I have met in this part of California. Both of these men are powers behind the throne in business and the higher planes of politics. They received me very graciously and listened with interest, asking easy questions, after Hubble introduced the subject of earthquakes." Wood went on to say, "They advised me against general public discussion of the subject—assuring me that it would array various interests and influences against my program which otherwise might be well disposed. They counseled rather a quiet, persistent campaign of education directed especially to the attention of individuals and small groups of influential men. This," Wood added, "is my own judgment emphatically."

In the same October 8 letter, Wood informed Day that there had been promising developments in the quest to find a suitable site in Pasadena to install monitoring instruments and establish a local laboratory. "A wealthy and philanthropically inclined citizen of Pasadena," Wood wrote, had expressed interest in donating a parcel of land in the San Rafael Hills west of Pasadena, land that Wood described as very suitable for a laboratory. Wood did not in his letter connect the dots between the advice offered by local business leaders and the potential largess from a member of the same community. In the end the Pasadena Seismological Laboratory was established on a three-acre parcel that Caltech purchased from a local businessman in 1925, and it is not clear if there was any connection between this purchase and the property that Wood had written about. But by all indications the message that community leaders communicated to Wood in 1923 was heard, loud and clear.

As efforts moved forward, quietly, in Southern California through the early 1920s, Bailey Willis increasingly stepped onto center stage. From the time he arrived at Stanford University, he had enthusiastically followed and then supported Branner's push for hazard mitigation and public awareness. When Branner died in 1916, the full mantle fell naturally to Willis. He became increasingly active in the organization Branner had worked so hard to establish, the Seismological Society of America. To a far greater degree than the presidents who served between Branner in 1911 and himself, Willis embraced the role—and the new stage—with his signature energy and enthusiasm. He teamed up with Harry Wood to compile a first-ever fault map of California, a project motivated in no small part by a desire to support the society and its causes. To compile the map, Wood benefited from Hill's ties to industry geologists, who by that time had compiled quite a bit of information about faults. By 1922 much work had been done, by academic and industry geologists, to map faults in Northern and Southern California. To fill in gaps in Central California, in the summer of 1922 Willis arranged for his son Robin to be paid four hundred seventy dollars to undertake a rapid survey of faults in that region. By September of that year, the map was completed to Willis's satisfaction. Willis and others then used the map, published by the *Bulletin of the Seismological Society* in 1923, to advertise and drum up support for the society, which still struggled to find secure financial footing.

As the Carnegie Institution program gathered speed and took flight, Hill's involvement with earthquake monitoring and the conversation about risk remained peripheral. Like Willis, Hill was in his sixties at this time and, like Willis, his professional interests had never focused on the study of earthquakes per se. By personal and professional happenstance it had fallen to Hill—the man who had long downplayed the hazard associated with earthquakes—to characterize the large-scale geological structures of Southern California and help compile a map of the region's fault lines. Both Hill and Willis might have been minor players in the opening acts of the great quake debate, but acts to come—including one in which the earth itself played a starring role—was about to nudge them both toward center stage.

CHAPTER 7

AT THE EPICENTER

> Suggest something about the earthquake you have
> heard about and immediately you feel like a worm.
> "Earthquakes, huh!—why, we haven't had one
> since 18—, and then it didn't amount to anything,
> anyway. Look at your storms back East."
> —*Virginia Hugill, letter to the* Los Angeles Times,
> *June 14, 1925*

In 1925, Bailey Willis remained intellectually active and physically energetic. He was still devoted to his children and wife, to whom he continued to write what can fairly be called love letters, their effusiveness and affection undimmed by passing years. The few photos of Margaret among Willis's papers at the Huntington Library reveal her to have been rather plain in appearance, and a woman not inclined to smile brightly for the camera. But her letters to her husband reveal unfailing good cheer and adoration. Throughout his life, Willis also remained close with his family, making regular visits to Morristown, New Jersey, to see his daughter Hope, his sister Edith, who had raised her for a time, and their brother Grinnell, who had become a prominent businessman and civic leader.

Professionally, Willis worked as a consultant on a range of projects, traveling widely throughout the state and the world. "His retirement . . . left him free for many years of activity of his own initiative," wrote Eliot Blackwelder in a memorial following Willis's death in 1949. "In California his studies were concerned largely with various parts of the Coast Ranges." During this time his work contributed to a US Geological Survey program to produce so-called quadrangle maps for the entire state. He continued to campaign for earthquake risk mitigation and, as Branner had done before him, for the Seismological Society of America.

Both of Willis's sons had launched careers as petroleum geologists. Cornelius, known as Neal, earned his undergraduate degree from Stanford in 1920 and in 1923 joined Marland Oil Company, where he became the personal assistant to the director in charge of geology and research. In this position, Neal coordinated research on geophysical methods and made recommendations for its practical applications. In 1924 Neal co-authored an article proposing a new mechanism to account for the development of *en echelon* oil fields along a deep-seated lateral fault zone. It was a fundamental contribution to structural geology after his father's heart. As of 1925, Neal's work with Marland Oil had brought him to the Los Angeles area. Willis's letters to Margaret reveal that their younger son, Robin, known as Bob, struggled somewhat to establish both his personal and professional lives; he, too, had graduated from Stanford and as of 1925 continued to live with his parents.

On June 28, 1925, Willis and Bob took a train south to Santa Barbara. A letter to Margaret, written that same night, chronicled his activities that day. He and Bob had proceeded to the Sammarkand Hotel, an elegant lodging just northwest of downtown Santa Barbara, where he found Neal, Neal's boss, and "the other geologists." Marland Oil, it appears, had taken an interest in the Santa Barbara area, where oil had been discovered back in the 1890s by none other than Ralph Arnold, and where a deep test well had recently been drilled in the Summerland field, east of Santa Barbara. Willis did not elaborate on the nature of the meeting; in general, his letters to Margaret and others were sometimes circumspect about the details of work that he undertook as a consultant. Following the meeting at the Sammarkand, Bob remained in Santa Barbara, while Willis proceeded to the Miramar Hotel east of Santa Barbara. When he penned his nightly letter to Margaret that evening, he wrote on Miramar Hotel stationery.

Whatever Willis's plans for the following day had been, the earth had other ideas. About what happened at 6:44 the following morning, no speculation is needed: we have Willis's own detailed accounts. "The writer," he said in his article later published in the *Bulletin of the Seismological Society of America*, "was at the hotel Miramar, four and one-half miles east of Santa Barbara. Lying awake, he heard, as it seemed, a train approaching along the Southern Pacific tracks from the *east*, experienced such rapid vibrations as are produced by a train close at hand, and then felt the sharp jolt of the advancing wave of an earthquake. It came from the *west*. He was thrown sideways in that direction." To the modern seismological ear, this last bit of the account is perplexing: a jolt perceived as coming from the west, throwing the observer sideways in the same direction. Earthquakes are now known, however, to generate waves that move the ground in complex ways: observers often describe a wave as coming from a certain direction, which turns out to not be the true direction of origin.

Willis's later account continued: "Recognizing the meaning of the shock, he noted the approximate time (6:44 a.m.) and began to count seconds. He had reached fifteen when the movement stopped. In the meantime the bed was rotating in an anti-clockwise direction with sufficient energy to cause him to put out his hand to steady himself." That he had just experienced a significant earthquake, Willis had no doubt. But it stopped short of generating real alarm. His account continued, "Had the motion continued or increased materially in violence it would have become alarming." As it was, Willis and his companion, a civil engineer by the name of Fitzgerald with whom Willis had apparently planned to work that day, "dressed without haste, taking nineteen minutes, and in that interval there occurred six earthquake shocks including the first." As Willis dressed, he noticed that the shocks had had no discernible effect on the wood-frame building around him, except that "the nails pulled in and out as it swayed after the first shock." Beyond his room, however, the shaking had left its mark. "Going downstairs," Willis wrote in a letter to Harry Wood, "I was surprised to find people rather frightened, and to learn that the chimney had fallen on the roof above me." In the general vicinity of the hotel, he wrote in his later article, "All the chimneys were thrown down, and a garage, constructed of reinforced concrete frame with hollow tile filler walls, was slightly cracked in the gable."

Willis's published article then described the account from Mr. Nunn, the city manager of Santa Barbara, who experienced the earthquake on the mesa just west of downtown Santa Barbara. After feeling an initial jolt that threw him on his back, Nunn ran from his bedroom to the front lawn. "As I stepped from the front porch to a slightly sloping grass plot," Nunn reported, "I was thrown violently down. My wife and the gardener, who were on the opposite side of the house at the time, came through [the house]. In stepping from a tiled porch to the grass . . . both were thrown down." Along State Street, the main business corridor in downtown Santa Barbara, the shaking found easy targets in the blocks of unreinforced masonry (mostly brick) buildings. "At the instant that Mr. Nunn was thrown backward by the initial impulse," Willis wrote, "the foundations of the buildings on State Street were thrown northeastward and the fronts fell into the street." According to other eyewitness accounts, the weakest buildings collapsed quickly, while "more firmly bonded structures presumably held together until wrecked by the succeeding vibrations or by their own oscillations." A civil engineer driving down State Street at the time described feeling a blow from behind "as though some one had run into the rear of his car. Turning to look, he watched the San Marcos building fall." The four-story San Marcos building had been constructed of reinforced concrete, a more earthquake-resilient type of construction than unreinforced masonry; an L-shape design had, however, left it more vulnerable to damage than a simpler structure would have been. Willis took the description of a blow coming from the north as evidence that the earthquake movement had come from the north or northwest, but again, such inferences about shaking direction are now known to be unreliable. For this reason, Willis's interpretations agree in some but not all respects with modern interpretation of the observations.

The 1925 Santa Barbara earthquake, as it was inevitably called, left a dramatic swath of destruction along State Street, the city's main business corridor, as well as pockets of serious damage in other areas. Throughout the residential neighborhoods of Santa Barbara, small wood-frame structures withstood the shaking fairly well, but brick chimneys toppled throughout the city. The temblor took a human toll as well: thirteen people lost their lives as masonry buildings crumbled onto them.

News reports focused on Santa Barbara, the largest population center in the immediate area and apparently the hardest hit. But the temblor took a toll in neighboring smaller communities as well. In a letter dated July 10, 1925, a woman named Katharine Maiers wrote to Willis, describing the earthquake in Goleta. Willis later passed along the letter to his colleague Harry Wood. "There has been little in the papers," Maiers wrote, "about Goleta damage, for we are not exactly a town, but a little community six miles north of Santa Barbara on the State Highway." In Goleta, which by compass heading is slightly north of due west of downtown Santa Barbara but along a northbound highway, Maiers informed Willis that "[they] did not have a business section to lose, [but] the proportion of homes destroyed will, I believe, exceed Santa Barbara's." Maiers provided her own eyewitness account: "To give you an idea of the severity of the quake in this particular spot, I will say that I was knocked through a doorway, face forward, on a porch. I tried repeatedly to rise but in each instance was thrown flat on my face. Finally the earth gave a great lurch, which picked me up from where I lay, tossed me in the air, turned me over and landed me on my back in the yard four feet away from the porch."

Maiers did not know, or did not explain, what direction she thought the waves had come from, nor did she have any foundation to try to explain what she had experienced, beyond drawing the conclusion that she must have been very close to the earthquake. Her description, however, lends itself to straightforward interpretation: she appears to have felt strong shaking from the initial P (for both pressure and primary) waves generated by an earthquake approaching her location. To understand this, a bit of basic seismology is in order. Any earthquake nucleates from a point and propagates along a fault at a speed of about 1.6 to 1.9 miles (2.5–3.0 km) per second, or about eight times as fast as the speed of sound through the air. As the earthquake moves along the fault, it generates both P and S waves. The P waves, akin to sound waves traveling through the earth, are faster, moving through the earth at speeds of approximately 3 miles (5 km) per second. S stands for both "shear" and "secondary": these waves, which involve ripples of sideways motion, travel more slowly than P waves. Because P waves travel faster than the fault break itself, they get out ahead of the propagating rupture front. If an observer is very close to a fault and an earthquake begins on that fault some distance away, they will first feel P waves from the fault at some distance from their location. The actual fault motion at that location occurs some seconds later, when the earthquake rupture itself passes by. The "great lurch" that Maiers described was likely the passage of the rupture; the fact that it threw her into the air and off the porch suggests that she was indeed very close to the causative fault.

Following many early earthquakes, particularly insightful accounts are written by eyewitnesses who, while not scientists, are clearly keen observers of nature. Katharine Maiers was one such witness. Her letter continued, "Another thing I should like to mention is the size of the cracks in the earth in our locality. These are large, and in some places water and blue sand bubbled up through them; while in other places one side of the crack is lower than the other, looking as if a section of the surface had settled down somewhat." Although a surface break associated with the earthquake was never mapped, conceivably some of these cracks were actual offsets along the fault expressed at the surface. She further noted, "There is another suggestion that the center of the quake is near us, in the fact that the intonation coming before the quake is so loud here, sounding not far distant. It is not the rumble that I have often heard described, but sounded more like a huge dynamite blast in rock."

In her interpretation, Maiers presumably did not know what geologists understood by that time, that large earthquakes rupture some distance along a fault, and the strongest shaking does not indicate the actual origin point of the earthquake. Now estimated at magnitude 6.6, the Santa Barbara earthquake was large enough to have involved movement along a swath of fault some 15 miles (25 km) long. Combining Willis's account at the Miramar Hotel with Maiers's in Goleta, it appears that the earthquake initiated slightly to the east of Willis's location (the human ear being generally good at discerning the direction that sound comes from) and propagated to the west, directly through central Santa Barbara, continuing very close to Goleta before losing steam somewhere to the west.

At magnitude 6.6—a magnitude estimated later using seismic data recorded by seismometers around the globe—the 1925 Santa Barbara earthquake was among the largest early-twentieth-century temblors in California. It is now believed to have involved motion on several abutting faults, one of which runs nearly through the heart of a vibrant business district. Bailey Willis had, moreover, been the man on the spot, predeployed, as it were, to witness the beginning of the earthquake firsthand. Inevitably, news reports surfaced that he had predicted the earthquake. This lore stemmed in part from comments Willis had made in speeches in 1923 regarding new data purporting to show that strain in the Santa Barbara–San Luis Obispo region was building up at an alarming rate. The data in question came from triangulation surveys conducted by the US Coast and Geodetic Survey. Most faults, including the San Andreas as well as most other faults in California, remain locked until they move in the abrupt lurch of an earthquake. But the locked zone is narrow; away from the fault, the two sides continue to move steadily. For the San Andreas, we now know that the rate of motion is about as fast as fingernails grow. If one painted a straight line across the fault and continued it for many miles on either side, over time the far ends of the line would move sideways relative to one another, bending only in a relatively narrow zone on either side of the fault. In lieu of that straight line, repeated surveying campaigns, which measure precisely the distance between a collection of fixed benchmarks, can detect this warping.

In 1923 initial analysis of US Coast and Geodetic Survey data indicated that a lot of warping had happened in Southern California in just the previous thirty years. This result got the attention of the scientific community, and Willis mentioned it in talks he gave to local civic groups.

Although the media coverage was not extensive, some stories appeared in newspapers. In December 1923, for example, the *Palo Alto Times* carried a front-page story, "New Quake Coming in about 15 Years, Rotary Club Told." The article described the maps that Willis had shown in his talk, illustrating the movement of benchmarks. Between this and his front-row seat to the Santa Barbara earthquake, the lore was easy to believe.

After the Santa Barbara earthquake struck, Willis appears to have not taken undue pains to refute this misperception, at least not publicly. Privately he did explain to colleagues that other business, not the expectation of an imminent earthquake, had brought him to the Miramar Hotel, although he did not elaborate on the details.

In the immediate aftermath of the earthquake, a geologist with keen interests in earthquakes might have remained in the Santa Barbara area to investigate the geologic effects of the temblor he had witnessed—for example, spending more time looking for clues about which fault(s) had produced the earthquake. Such work would have been of value to later

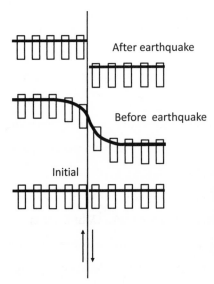

Illustration of how strain builds along the San Andreas Fault as the two sides move sideways and is released during an earthquake: (bottom) fence built straight across a fault; (middle) fence warped as strain builds along the fault; (top) fence broken after earthquake.

generations of scientists. But Willis wasted little time in returning home. As he noted in his June 28 letter to Margaret, he and Arthur Day had a commitment to attend a meeting of the San Francisco Chamber of Commerce on Thursday, July 2. Thus did the man who had been virtually at the epicenter of the Santa Barbara earthquake team up with Day to address a gathering of 150 businessmen at the San Francisco Chamber of Commerce, just three days after the earthquake. Day later reported that he and Willis had spoken plainly about the need for earthquake risk mitigation. A month later, Willis again addressed a group of influential business leaders in San Francisco. In this speech, he assured the audience that he was not opposed to growth but, rather, he argued that long-term growth required rigid and effectively enforced building codes. By 1925, one notes that earthquake awareness and economic growth tended to be painted as either-or propositions.

In a public arena as well as behind the scenes, Willis's crusade continued, with more speeches, interviews, and articles. Both he and Wood wrote articles for a special issue of the *Bulletin of the Allied Architects Association of Los Angeles*, put together a month after the earthquake. Willis's lead-off article described the geological faults in the Santa Barbara area and the "unnatural faults" in substandard building construction that had contributed so significantly to the damage.

Wood's article mentioned the buildup of strain that Willis had spoken about two years earlier. Combined with the fact that the southern San Andreas Fault had not produced a significant earthquake since 1857—and a sizable earthquake was known to have also hit the Southland in 1812— the buildup of 24 feet (7.3 m) of strain across the region seemed ominous. Harry Wood might have been a naturally prudent man, but privately he shared Willis's concerns about a possible future large earthquake, if not his definite statements regarding likely time frame. In Wood's view, the spate of moderately large earthquakes in Southern California since 1915 were an ominous indication that strain in the region was building toward a critical level. This view of an earthquake cycle, in which regional activity increases in advance of a large earthquake, still finds some support among the modern earthquake science community. Although statistical seismologists have argued fairly convincingly in recent years that this idea is not well supported by observations, the view was not unreasonable in the early twentieth century, and the indication of recent strain accrual further stoked Wood's concerns. It was not in Wood's nature to take his concerns to the media and the public, especially not having been specifically cautioned against doing so, but in the aftermath of the Santa Barbara earthquake he was willing to express at least some concern in a professional article. The rest of the mostly brief articles in the special issue were written by architects, engineers, and general contractors, some of whom made the argument that ordinary construction, if of sufficiently good quality, was enough to make a building earthquake-safe. Other articles argued that specialized design, beyond ordinary sound construction, was necessary to ensure that buildings could withstand earthquake shaking.

In addition to writing his own technical articles, Willis continued to attend private meetings and make public statements in media interviews. For Willis, Wood, and other scientists, but Willis especially, the earthquake presented an opportunity. Willis issued a press release of his own, circulated widely by the Associated Press. The version circulated by the AP included an introductory paragraph that by some accounts had been inserted by the publicity department of San Francisco–based Californians, Inc., making the claim that Willis had proclaimed Santa Barbara "now [be] one of the safest places in the United States so far as earthquakes are concerned." Whether or not Willis himself wrote the claim, it did echo a common view shared by many earthquake scientists in the early twentieth century, that a large earthquake effectively "immunized" a region for some time to come. Willis's article, which emphasized the importance of better building practices, reached many readers throughout and beyond the state.

In the immediate aftermath of the 1925 earthquake, while Willis crusaded, Robert Hill remained out of the public eye. Newspaper articles published after the Santa Barbara earthquake quoted a number of experts in addition to Willis, Hill not among them. Although he had worked for four years on a monograph summarizing his fieldwork in Southern California, the work was never published. According to Carl-Henry Geschwind's account, the report was "too rambling and incoherent." According to Hill, by 1917 the first world war had left the USGS too preoccupied with other matters to bring a lengthy report to fruition. As was often the case in Hill's life, the truth most likely lay somewhere in between. Hill's talents for understanding geology and penchant for putting together extensive geological reports were not accompanied by the wherewithal to organize lengthy reports without substantial editorial assistance. The onetime newspaperman could spell words correctly and write grammatically correct sentences; in later life especially, however, he appears to have struggled to string those sentences together when reports stretched to hundreds of pages. One does note, however, that, as had been the case with his earlier work in Texas, Hill's extensive mapping in California was useful to the USGS and others in many ways, even in the absence of a comprehensive report. Hill also fundamentally simply bit off more than he could chew, so to speak, bringing his keen geological acumen to the task of mapping enormous and complex regions, leaving himself the impossible job of writing encyclopedic reports almost single-handedly. His complaints about a lack of secretarial and editorial support may have been tinged with a trademark tone of petulance and rooted in his own style of working, but were not altogether unfounded. Few if any other geologists would have taken on as much work as Hill had in California and set out to write up a comprehensive report on their own. Perhaps by this time, USGS managers knew better than to expect that Hill's maps would be accompanied by a monograph.

The early 1920s were not in general good years for Hill. His health declined during this time, with several heart attacks and a diagnosis of myocarditis, requiring periods of hospitalization and recuperation. Letters to his old friend Ellis Shuler ran toward maudlin. "I know that my heart is in terrible shape," he wrote in 1923, "and that I may never write you again." Like Willis, he supported himself during these years by working as a consultant. In 1921 he was among the experts chosen by the state of Texas to offer expert testimony in a legal dispute between the states of Oklahoma and Texas regarding ownership of the Big Bend area, about 100 miles (160 km) northwest of Dallas. At the heart of the case was the question of whether the Red River, which defines the state line, had migrated since the border between Oklahoma and Texas was established in 1821 by a treaty between the United States and Spain. Since oil had recently been discovered in the Big Bend region, the precise position of the border was a matter of some interest to both states. Hill argued that, while young rivers meander, "rivers like men become fixed in their habits with the approach of old age." There was, he concluded, no geological evidence that the river had shifted its course in the single century since the treaty had been signed. In early 1923, the Texas Supreme Court issued its ruling in the Red River case, giving Texas ownership of 450,000 acres of river valley land and about 90 percent of the oil wells therein. Hill's role in the case bolstered Texas's pride in claiming him as a native son—a claim that Tennessee generally did not contest.

By the end of 1924, however, with his younger daughter Jean in boarding school and a dwindling bank account, Hill's physical ailments had left him struggling to work. The Santa Barbara earthquake did not, of course, escape his attention entirely. Six days after the earthquake he wrote to Ellis Shuler, "We had a hideous earthquake out here at Santa Barbara lately. It was by far the worst one they have yet had in California in the memory of man, but owing to the fact that there were no fires the damage to life and property was not as serious as was that of the San Francisco earthquake."

Hill's health did improve in 1925, and he accepted a one-year contract from the Los Angeles Museum to write a report about the geology of the Los Angeles region. Hill, author of several exhaustive geological monographs, whose work on a USGS monograph on California had stalled out, saw it as an opportunity to produce a "great book," one that would present "the story of the Los Angeles Region during the Quaternary Period—and a wonderful story it is." The contract apparently provided not inconsiderable financial support: by Hill's account in mid-December 1925, the project had kept him "monomaniacally employed for the last six months." He began 1926 on an optimistic note, writing in his diary on New Year's Day that he was starting the year "with a clear conscience, good health and more happiness than for years. M. is better and like her old self. I have learned much about life and philosophy in the past six months and have some good friends." Ralph Arnold and other members of the Los Angeles Chamber of Commerce's earthquake committee, which Hill had joined by this time, were among those whom he counted as close friends.

This assault on his book, however, never came to fruition. By 1926 he had completed an 805-page draft, but before Hill could finish the project, illness again landed him in the hospital for seven weeks. His health improved to the point that he accepted a temporary appointment at the University of California, Los Angeles. Released from the hospital only days before the start of classes, he worried that he would not be able to teach. "They hustled me out of the hospital two or three weeks before I should have left it," he wrote to Shuler on October 24, 1926, "in order that I might fulfill a contract I had entered to take a teaching place for a year." In the end, however, he showed up for the first day of classes and lectured for three full hours. "It uses up all of my pep and strength," he wrote to Shuler, "and I come home after 1 o'clock and throw myself on the bed and am not much good for the rest of the day." Exhaustion notwithstanding, the lively and progressive university community, unlike the stodgy academic milieu he had found in Austin years earlier, proved to be just what the doctor ordered, so to speak. He even found energy to take his students, including more than a few women, on field trips, although less frequently than he had in his bouncier University of Texas days. By the end of 1926, his health and spirits again began to rebound.

The 1925 Santa Barbara earthquake itself, so pivotal for Willis and the great quake debate, does not appear to have been more than a passing blip on Robert Hill's radar screen. A combination of illness and financial concerns pulled him off of the stage entirely for this act, focusing his personal and professional interests elsewhere. But as his health and spirits rebounded, he landed on a trajectory that would soon bring him into a full head-on collision with his old nemesis.

CHAPTER 8

THE PREDICTION

> Occasionally a professional man who has a good
> reputation in other fields is responsible for errone-
> ous statements about earthquake occurrence and
> earthquake prediction. Even good geologists have
> been known to fall into such errors.
> —*Charles F. Richter*

Bailey Willis and his colleagues had some real successes in their campaigns in the immediate aftermath of the Santa Barbara earthquake. The community of earthquake professionals had grown by this time. Thanks in no small part to Willis's own efforts, the Seismological Society of America was on sounder footing than it had been in earlier years. The statewide fault map that Willis and Wood had teamed up to compile and publish represented a key product for several reasons, of interest both to those interested in earthquake hazard and to those interested in understanding geological structures, including major faults, to guide the search for oil and gas.

After the 1925 Santa Barbara earthquake, buoyed by the efforts of earthquake professionals and the dramatic demonstration of earthquake risk, the cities of Santa Barbara and Palo Alto incorporated seismic provisions into their local building code for the first time, awareness of earthquake hazard increased throughout the state, and in the Bay Area, some in the business community lent support to scientific and engineering research. Gains were, however, limited. Business leaders and city boosters also sprang into action quickly following the Santa Barbara earthquake; within days of the earthquake, representatives from several prominent business organizations met at the Los Angeles Chamber of Commerce to formulate plans. A nationally syndicated columnist for the Hearst newspaper chain, for example, was convinced to describe the earthquake as a "local isolated slip with no recurrence probable for many years," adding that "reconstruction has already started and the local situation will return to normal promptly." On July 2, 1925, the *Los Angeles Times* published a short article with the headline "Splendid, Santa Barbara." This article noted that damage had been light, and "more than thirty American cities have suffered a greater loss by fire or cyclone, some of them twenty times as great." The article continued in the same reassuring vein: "So far as earthquakes are concerned, Santa Barbara is today about the safest community in the whole country; and this security can be counted on to continue for at least 100 years. We of the West are not so foolish as to fear earthquakes that have passed. They resemble measles, mumps, and certain other ailments that only afflict a person once in a lifetime."

The Los Angeles Chamber of Commerce went so far as to take its public relations campaign on the road. It dispatched Ralph Arnold, by this time head of the Southern California Branch of the Seismological Society of America, to make a tour of eastern cities, "largely to acquaint inhabitants with facts concerning the recent Santa Barbara shake." Arnold might have struck Harry Wood as an overly enthusiastic booster for earthquake monitoring just a few years earlier. By 1925, however,

Arnold's interests appear to have been more aligned with the business community. In a talk at Columbia University, Arnold told the audience that the Santa Barbara earthquake had not been so bad, that "jerry-building was responsible for the loss of life and . . . ordinary prudence in building construction would have not only prevented loss of life, but would have restricted property damage to a few fallen chimneys." He emphasized that subsurface strata in California were similar to that in other parts of the country and that earthquake hazard in California was no different from that in other regions. He talked about the 1811–12 New Madrid earthquakes in the central United States and the 1886 Charleston, South Carolina, earthquake, the latter not yet faded from living memory.

At the time, some of Arnold's points were not altogether unreasonable. Before the 1935 introduction of the magnitude scale, scientists had no way to measure the intrinsic size of an earthquake. The distribution of damage and felt shaking was taken, not unreasonably, as an indication of an earthquake's strength. The far-reaching effects of the 1886 Charleston temblor, the felt shaking from which reached well into the Midwest, dwarfed the documented effects of any known earthquake in California. Even the 1906 San Francisco earthquake, which left a dramatic scar along the San Andreas Fault, was barely felt in Nevada. What scientists know now but didn't know then is that the distribution of shaking depends only partly on the actual size of the earthquake. In central and eastern North America, waves travel far more efficiently in the relatively homogenous old crust than they do in the more shattered crust in California. We now know that neither the largest 1811–12 New Madrid earthquakes nor the 1886 Charleston earthquake were as big as the largest historical earthquakes in California. One hundred years ago, however, scientists reasonably concluded otherwise.

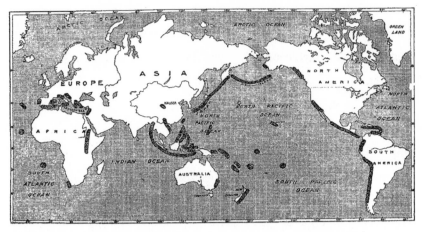

Global earthquake zones identified by 1923. (*New York Times*, August 9, 1923)

Interestingly, during the months before the Santa Barbara earthquake, the earth had done its own part to underscore Arnold's point. On February 28, 1925, an earthquake with estimated magnitude of 6.2 struck Quebec, generating shaking that was felt as far south as Virginia. This event sparked media interest in earthquakes in the northeastern United States. A small flurry of articles appeared in prominent East Coast newspapers, including the *New York Times*, talking about earthquakes in eastern North America and sometimes downplaying the severity of hazard in California. Just a couple of years earlier, the *New York Times* had published a map indicating that a ribbon running through California was the only significant earthquake zone in the contiguous United States. On August 1, 1925, after the Santa Barbara earthquake struck, the *Times* published another map indicating with shading where earthquakes had occurred in the past year; a small swath of shading in Southern California was dwarfed by larger shaded regions extending through much of the northeastern United States and into the Midwest, as well as parts of Oklahoma, Texas, and Mexico. As it happened, Ralph Arnold spent quite a bit of time in New

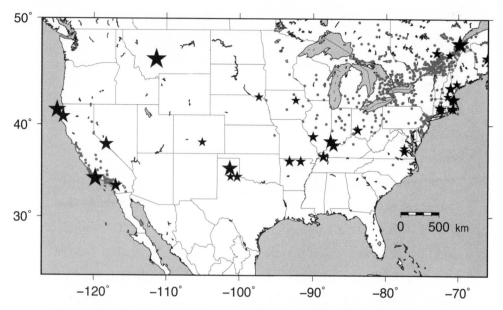

Recorded earthquakes in 1925 (stars scaled by magnitude); locations where the 1925 Santa Barbara earthquake was reported felt in California (dots); locations where a similar-sized earthquake in Quebec was reported felt on the East Coast (dots).

York in early 1925, with occasional trips to Washington, DC, and would have been well positioned to shape media coverage of earthquakes during this time. But the earth itself had underscored one of his key points: judging (reasonably) from the extent of shaking, the 1925 Quebec earthquake had been far more portentous than the 1925 Santa Barbara earthquake.

By the time the Santa Barbara earthquake struck, the population of Los Angeles County had passed the one million mark, about a third of the state population. The continuing oil boom fueled population growth in the 1920s, with the discovery of a series of oil fields along the Newport-Inglewood Fault, starting with the Huntington Beach field in 1920. By 1930 the county population would grow to 2.2 million, almost 40 percent of the total state population. In the aftermath of the Santa Barbara earthquake, the public relations campaign led by city boosters was a well-oiled machine, drawing on a wide-ranging web of highly influential and often wealthy contacts with a stake in the continued economic success of the region. And while the damage in 1925 outstripped that in the 1918 San Jacinto and 1920 Inglewood earthquakes, there were still valid points to be made about biased reporting. Along with many of his colleagues, Willis himself noted that damage had been restricted to vulnerable masonry construction.

Statements made by city boosters after the 1925 Santa Barbara earthquake echoed those made following the 1918 San Jacinto earthquake, but were disseminated via a more coordinated campaign. Eventually, some in Santa Barbara complained that the effects had been downplayed to the point that they struggled to get much-needed assistance. In Los Angeles, it remained possible to paint earthquakes as an inconsequential local concern. Shaking had certainly been felt throughout the greater Los Angeles metropolitan region on the morning of June 29, 1925, but with rare exception, damage extended no farther east-southeast than the coastal community of Ventura, still comfortably distant from Los Angeles. By 1925 people in the Los Angeles area had experienced both the 1918 San Jacinto and 1925 Santa Barbara earthquakes; waves from both earthquakes gave the Los Angeles area a solid shaking, but, with rare exception, the waves lost their destructive punch by the time they reached the rapidly growing metropolis. The local but smaller Inglewood earthquake of 1920 had also been soon forgotten and not remembered by the thousands of Angelenos who arrived in the area after the earthquake struck. Through the early 1920s, Southern Californians could believe the line that business interests spun: yes, earthquakes sometimes occurred in Southern California, but they were little more than a passing nuisance. Even earthquake-stricken Santa Barbara bounced back, the city's population growing from 31,744 in 1925 to 32,274 in 1926, with a 12 percent increase in total property valuation from 1924 to 1926.

In Southern California, the roaring twenties continued to roar. New oil fields were discovered, and new cities sprang up alongside them: Huntington Beach, Long Beach, Santa Fe Springs, and more. Today we count them among the sea of contiguous towns within the sprawling patchwork quilt that is the greater Los Angeles metropolitan region. In their day, they were separate cities—oil towns, all. The efforts of city boosters did their part to erase concerns about earthquakes, but in any time and place, the news cycle inevitably moves on after a damaging earthquake. Back on the East Coast, the recent 1925 Quebec earthquake had obligingly underscored one of the key arguments made by city boosters: that earthquake hazard in California was not high compared to other parts of the country. City boosters had not, of course, missed the opportunity this earthquake had provided. About a month after the earthquake in Quebec, Ralph Arnold wrote to a member of the Los Angeles Chamber of Commerce, including a July 24 article in the *New York Times* in which Arnold himself was quoted as saying "The East or the Mississippi Valley is quite as likely to be the locality of the next great earthquake as the Pacific Coast is." In his letter Arnold noted, "I think it is well written about the type of information we want to broadcast." In the same letter, Arnold wrote that, in his own speech, he "blamed all of the deaths [in the Santa Barbara earthquake] to faulty construction."

For a confluence of reasons, whatever traction Willis and his colleagues had found in the aftermath of the Santa Barbara earthquake lost steam before 1925 drew to a close. Overall, in the great quake debate, the 1925 round appeared to go to business interests, who had not only their well-connected machine to draw from by that time but also, not inconsequentially, the fact that the earthquake had struck some distance from the greater Los Angeles region.

The scientific community, however, remained galvanized. In August of that year, Caltech's Robert Millikan established the institute's own Southern California Council for Earthquake Protection. The council held initial meetings that month, extending invitations to a number of prominent geologists, including Wood, Willis, and the one geologist who continued to swim equally comfortably in academic and industry waters, the ubiquitous Ralph Arnold. As Millikan's earlier committee had done back in 1921, the council pressed for development of an integrated program to attack the earthquake problem. Progress on risk mitigation, Millikan argued, could be made only via a coordinated program involving a number of key agencies, including Caltech, the US Geological Survey, and the US Coast and Geodetic Survey. In keeping with the modus operandi of the Pasadena Seismological Laboratory during its early years, these efforts moved forward quietly, with no concerted public relations campaign.

Whatever understanding had been worked out between Wood and local business leaders in Southern California in 1923, Bailey Willis, comfortably distant in the San Francisco Bay Area, never signed on to the deal. In the words of historian Carl-Henry Geschwind, by the end of 1925, "[Willis] decided to embark on a new strategy: he would scare Californians by not only pointing backward at the recent seismic destruction but also predicting that a catastrophic earthquake, much larger than the Santa Barbara shock, would soon strike southern California."

After the worrisome US Coast and Geodetic Survey result became known in 1923, scientists like Wood avoided making overly alarmist public statements. In addition to having been specifically cautioned about public statements, Wood knew that analysis of early triangulation data could be an imprecise science. The results in question were in fact preliminary, estimated from triangulation measurements that had not been properly connected to the larger regional triangulation survey. Moreover, scientists would have been reluctant to go public with a scientific result that raised some concern in scientific circles but did not support a specific forecast, let alone a precise prediction. As Wood himself had written cogently in his 1916 article, "At what time future shocks will occur we do not know, especially in any precise way; but we do know that since 1769 no half-century has passed without the occurrence of at least one great earthquake in this region. The average points conclusively to greater expectable frequency. However, close prediction of the occurrence as of day or hour, even of month or year, cannot even be approximated as yet." With the US Coast and Geodetic Survey result, even a prudent scientist like Wood faced a dilemma that remains familiar to present-day earthquake professionals. On one hand, an apparently sound scientific result raised concern within professional circles, but on the other hand, the result was clearly uncertain to some extent, and in any case the result did not imply any specific time frame or even a quantifiable statistical forecast. Both Wood and Willis did know, however, that a regional buildup of strain—if it existed—would be released eventually in a large earthquake and that the relatively modest Santa Barbara earthquake had not been that earthquake.

Unlike Wood, Willis had spoken publicly, to a limited extent, about the buildup of strain before the Santa Barbara earthquake struck. Before 1925 drew to a close, he latched more squarely onto the US Coast and Geodetic Survey result. Pointing to the result, he told *Daily Palo Alto* readers (among others) in November 1925 that a large earthquake in the Southland was nigh. "No one knows whether it will be one year or ten years before a severe earthquake comes," he said, "but when it does come it will come suddenly, and those who are not prepared will suffer."

In the earthquake business, then as now, there can be a vanishingly fine line between saying enough to get people to take earthquake hazard seriously and being overly alarmist. In either direction away from that line, unfortunate things happen. In one direction, the public and decision makers ignore warnings altogether; in the other, people might panic or fail to take action ("We're all doomed, what's the point?"), and there starts to be the danger of crying wolf too often. Now, as in Willis's day, some individuals flirt more closely than others with that fine line. In the statements he made in November 1925, Willis did more than flirt: *"No one knows whether it will be one year or ten years."* In fact, then as now, no one knew whether it would be one year or ten years or a hundred years. The modern reader can view Willis's words with the benefit of almost a century of hindsight. Damaging earthquakes did strike the greater Los Angeles area, including not only the one that would (mostly) put an end to the great quake debate, but also a moderately damaging earthquake near Whittier in 1987 and a pair of larger temblors that struck the San Fernando valley, not too far north of Los Angeles, in 1971 and 1994. Yet the earthquake that Willis, and some of his colleagues, warned about, a great earthquake in Southern California rivaling the 1906 San Francisco earthquake, did not occur within three years or ten years or ninety years of 1925.

When a scientist dances with or oversteps that vanishingly fine line, the media, which is always called on to translate scientists' statements into actual English, invariably and not altogether unreasonably drop finely nuanced qualifications. Headlines are, moreover, designed to grab the reader's attention, with less room for subtlety than a modern Twitter tweet. Willis's words soon found their way into the national media; the *New York Times*, for example, published an article titled "Prof. Willis Predicts Los Angeles Tremors": "Los Angeles, or its immediate vicinity," the article began, "will experience a severe earthquake, probably more violent than that at San Francisco in 1906, in one to ten years, Dr. Bailey Willis . . . said here last night." The article went on to repeat the lore that Willis had "stated three years ago that Santa Barbara would feel severe earth tremors, a prophecy that was fulfilled in the past Summer."

Over the years following the 1906 San Francisco earthquake, the great quake debate in Southern California had, at various times, simmered and stewed and flared. With Willis's public words in late 1925, it exploded. If Willis's own statement stopped just barely short of an actual prediction, he never publicly refuted the supposed prediction, nor did he ever take pains to refute the enduring misperception that he had predicted the Santa Barbara earthquake. Rather, his crusade continued on an increasingly visible stage, with nearly immediate effects. In a letter dated November 30, 1925, the general manager of the Los Angeles Chamber of Commerce wrote a letter to Ralph Arnold: "I send you herewith a letter from Dr. C. A. Williams of Los Angeles with a marked article by Professor Bailey Willis, citing the fact that in the next ten years Los Angeles is due for a severe earthquake. It may be within the realm of sciences to prognosticate these occurrences, but a careful scientist seldom prognosticates. I would like to have your frank opinion of this matter." Written to a geologist who was prominent in local industry and political circles, the letter reads less like a query than a call to action.

Although effective earthquake risk mitigation involves a number of professional communities—builders, architects, public officials, and more—the sparks from Willis's statement caught fire among the one industry for whom literal fire is a bread-and-butter concern: the insurance industry. Before 1925, earthquake insurance had not been a big business in the United States. Believing the assurances of boosters that earthquakes were not a serious concern in Southern California, the insurance industry actively pushed earthquake insurance in California in the months following the Santa Barbara earthquake. Insurers moreover convinced a number of banks to require earthquake insurance before a mortgage could be taken out on a commercial building. In 1925 alone, total earthquake insurance premiums collected in California nearly doubled the total premiums collected between 1906 and 1924. With talk suddenly in the air, from a leading scientist no less, of an imminent earthquake, insurance executives and banks, concerned about potential liabilities, took note.

This particular business community was, to a large extent, concentrated not in Los Angeles but, rather, back East. In May 1926, the National Board of Fire Underwriters in New York City invited Willis to give a keynote address at their meeting. In this speech he again stopped just barely short of a definite prediction with the words "I regard as probable that in Southern California there will be a severe shock which is more likely to come in three years than in ten, and more likely to come in five years than in three." It is not clear why Willis zeroed in on five years as the most likely time frame. Conceivably the time frame was the result of calculations he had done, but it might have simply been a pragmatic choice, close enough to be a real and present danger, far enough away that effective mitigation steps could still be taken.

Around the same time, the National Board of Fire Underwriters called on a team of engineers to survey some 2,700 commercial buildings in San Francisco, Oakland, Los Angeles, and San Diego. This survey revealed that, in the engineers' opinion, most commercial buildings would suffer severe damage in a major earthquake. The insurance industry realized it faced potentially enormous exposure, if Willis's predicted earthquake occurred. Insurers raised rates for earthquake insurance on commercial buildings, nearly doubling the rate for steel-frame buildings and nearly tripling the rate on reinforced concrete frame buildings as well as the most common type of construction at that time, unreinforced masonry. The industry had raised rates somewhat in the aftermath of the Santa Barbara earthquakes; in total, rates increased five- to tenfold, and in some cases even more, over just a few years.

The rapid insurance rate increases hit building owners and developers hard. In response, some in the business community joined the call for improving resilience of buildings. An insurance committee set up by the California Bankers' Association recommended that mortgages be approved only after a building had been inspected by a qualified engineer and deemed to meet certain standards. Another development association joined the call for earthquake provisions in the statewide building code. It had been Willis's gambit to spark rate increases in the hopes that the insurance industry would join the crusade for risk mitigation, and the gambit worked, to a point. But the sharply increased rates also angered local real estate developers and speculators. This group, drawing from abundant resources and political connections, wasted no time in fighting back. They took aim not only at Bailey Willis and his prediction but also, it seemed, at the larger cause of risk reduction as well.

As Willis's crusade brought the cause of risk mitigation out of the woodwork, some in the scientific community were drawn out of their shells. In the spring of 1927, two California Institute of Technology professors, John Buwalda and Romeo Martel, began work on their own public statement. The planned statement did not endorse the prediction per se, but laid out the authors' own concerns about a future large earthquake in Southern California. This plan drew immediate fire. After learning that the statement was in the works, Henry Robinson, a local city booster and trustee of Caltech, wrote in the spring of 1927 to the executive director of that institution, "I wonder if you have any idea how much damage this loose talk of these two men is doing to [property] values. . . . You can hardly appreciate how serious the situation is here and if we . . . cannot stop their talk about the earthquake problem I for one am going to see what I can do about stopping the whole seismological game." In April 1927, the warning, from a man who was by that time on the board of directors of the California Development Association, was no idle threat. With hard-won support from the Carnegie Institution and Caltech itself, the Pasadena Seismological Laboratory had just barely gotten off the ground. A fledgling network of Wood-Anderson seismometers had been built and was starting to be installed. Caltech had constructed a new building to house the laboratory in the foothills of the San Rafael Mountains, a short distance from the Caltech campus in Pasadena, allowing Harry Wood and his colleagues to move out of borrowed lab space at Mount Wilson. An agreement between Caltech and the Carnegie Institution gave Carnegie the exclusive right to conduct research in the building. To help analyze the new data, the Seismo Lab, as it has been known for generations of scientists, brought in a bright young researcher who had just earned his PhD in physics at Caltech: twenty-seven-year-old Charles Richter. Richter, who hoped to find a position in modern atomic physics, accepted the position with a measure of reluctance, intending, although it didn't turn out that way, that the job be only a temporary one.

The partnership between Carnegie and Caltech soon gave rise to skirmishes over ownership and control of the Seismo Lab's activities. The skirmishes continued until 1941, when Carnegie exited the scene, leaving Harry Wood as the sole remaining researcher in their employ. In the late 1920s, the Seismo Lab operated as a partnership, and funding for the new enterprise remained of paramount concern. Not long after the Seismo Lab hired Richter and began to work with the seismic data, Wood knew that he wanted to keep Richter on board but had to talk to Arthur Day at Carnegie about securing funds to keep Richter on the payroll permanently. Throughout the 1920s, while Carnegie remained committed to supporting the earthquake program in Southern California, the lab grappled with general financial pressures that forced cutbacks across its portfolio of activities.

The California Institute of Technology, meanwhile, had its own concerns. The university is today a world-class research and teaching institution, supported by a large endowment and research grants, the epicenter of the intellectually vibrant city of Pasadena. Caltech did not get to where it is today by running afoul of local interests. In the early years especially, Millikan was wary of government support for research, preferring to look instead for support from the private sector. Not surprisingly, the relationship between the university and business leaders appears to have been warm from the start. In November 1920, Ralph Arnold had written to the acting president of Caltech, expressing his keen interest in seeing that "a strong geologic department is built up somewhere in southern California, and I think your institution is the one to take the lead." Arnold went on to say, "I have in mind certain specialities, such as seismology and petroleum geology, that should be featured in such a department, and in order to do this it will probably be advisable to choose a man with great care. Any information that you can furnish along this line will be held confidential and will be greatly appreciated."

It is not clear how much involvement Arnold or others had in hiring decisions, but five years later, when Caltech President Robert Millikan invited John Buwalda to chair the newly formed Geology Division, they chose a scientist with not only impeccable academic credentials but also a geologist with close ties to industry. At Caltech, Buwalda supplemented his academic salary with private work undertaken as a consulting geologist. Indeed, Buwalda had a good relationship with Robert Hill himself.

In 1928 Buwalda wrote to Hill, "You may take some satisfaction in realizing that a very considerable part of my interest in southern California is due to the very interesting journey which I took with you quite a number of years ago in the course of which we examined a number of significant localities." A close relationship between Millikan and Arnold, meanwhile, was in evidence when the former man invited the latter to join the distinguished fact-finding committee that Millikan created in 1921 and the Southern California Council for Earthquake Protection he created in 1925.

By the time Robinson wrote his letter in 1927, the Carnegie Institution paid the salaries of Wood, Richter, and other Seismo Lab staff, but Caltech, increasingly involved with lab activities, had its own supporters and its own concerns. In response to Robinson's letter, Buwalda and Martel shelved their plan to issue a statement, and Harry Wood made no further public statements about earthquake hazards in Southern California. In general, when scientists from Wood to Willis to Hill spoke of the need for earthquake monitoring and research, they always framed the discussion in terms of the role that such activities have in improving safety. They further emphasized that properly constructed buildings would withstand even severe earthquake shaking, repeatedly stressing that damage in recent earthquakes, including Santa Barbara, had been limited to the most vulnerable construction. It was not too different from the line that Ralph Arnold had spun in his speaking tour following the Santa Barbara earthquake: earthquakes did not need to be feared; the risk could be managed. Whereas, however, the business community tended to emphasize that a significant earthquake effectively "immunized" a region from future earthquakes for decades if not centuries to come, the scientific community increasingly united behind the message that future and even larger earthquakes were expected in Southern California.

Although Willis was thus not alone in his concerns, and his goals were shared, he was more inclined than his colleagues to bring his concerns to a public stage. In the usual telling of the great quake debate, Willis emerges as a flawed hero, but a hero nonetheless: a lifelong advocate of Progressive causes, a champion of the Seismological Society of America, a tireless crusader for earthquake risk reduction. Whereas even Hill's close friends and colleagues acknowledged that he could be difficult, colleagues close to Willis chose other words to describe him: charismatic, charming, passionate, and tireless. When Willis went public with his prediction in late 1925, however, at least some suggested, at the time, that different motivations were at play. A short editorial printed by the *Santa Barbara Morning News* in early 1926 began, "One would judge that the doctor became somewhat spoiled at the period of our June disturbance last year through the publicity he garnered and that he is loathe to give up his place in the newspapers and magazines." The article went so far as to offer helpful advice: "We suggest that someone take the ebullient seismologist in tow and lead him into other pastures, if he must have publicity. He might have his photograph made with a California beach mermaid and even make the rotogravure sections and the moving picture magazines if he tried."

It perhaps attests to the ebullient seismologist's good humor that he did save the *Morning News* article among a sizable collection of other newspaper clippings in which his name appeared over the years. But did Bailey Willis have other reasons to seek the limelight? Did the adored little boy grow up to be a man who enjoyed being center stage? When approached by a reporter from the *Houston Post* back in 1898, did Willis try to tell the reporter that his traveling companion and colleague, Robert Hill, not himself, was the leading authority on Texas geology, so that Hill would be featured more prominently in the stories that were written? We don't know if he tried to deflect reporters' attentions to his more knowledgeable colleague, but we do know that, if he tried, he did not succeed.

We also know a bit about Willis's antics later in life, written by people who knew him. For example his longtime friend and Stanford colleague Payson Treat wrote a letter to Willis's daughter after her father's death: "I like to remember the way your father used to police the parking around the post office and bawl out the students who parked on the left side of the street. I would see a student wrongly parked, and observe your father approaching with the light stride of a catamount. I would wait for Act II, and a red-faced student pulling away from the kerb." Treat went on to recount another incident in which Willis confronted students, who had propped their feet on a balcony railing during a talk: "Your father was waiting for those young boors, and in a few well-chosen words he not only reflected on their upbringing but indicated how they might make amends for it." Treat himself most certainly applauded: "What a man!" he wrote. "We need him more than ever in this day of mass education." One can imagine, however, how Stanford students felt about the incidents that Treat described, what words some of them might have used to describe their ebullient professor.

Among Robert Hill's professional papers, archived at the DeGolyer Library at Southern Methodist University, a hand-scribbled, marginally legible note attached to a popular article about Bailey "Earthquake" Willis provides another glimpse into how Willis was regarded by some colleagues. "Look who's here!" the note begins, "Earthquake Willis! Back in geological survey days he was only Windy Willis! And that only to the irreverent small fry." The note continues, "An honestly descriptive heading of this ad, far over on the charitable side, would have been Blatherskite Willis at 90, still reminiscing at the mouth! Ever notice—some people remind you of animals? A hoss-faced man? A snake-eyed woman, etc.?

Knowing Willis and looking at his picture today, he suggests a half-[unintelligible], overage seavoyager bird—unsavory memory of meagerly dogs on a Georgia cotton plantation. (If too crude do forgive me. Polite language would not answer.)" The note presents a mystery: the article to which it is attached was published in *Newsweek* magazine in June 1947, six years after Hill's death. The note and article might have been filed with Hill's other (scant) correspondence pertaining to Willis by Hill's daughter, Justina, who kept possession of her father's papers for years after his death, with the intent to write his biography. Although the project never came to fruition, Hill's papers remained in Justina's possession until 1972–73, when she passed them along to Southern Methodist University, where Ellis Shuler had collected other material on Hill until the time of his death in 1954. It appears, then, that the note had either been sent to Shuler and later merged with the folder of Hill's correspondence pertaining to Willis or sent to and filed by Justina herself in 1947. In either case the note reveals some clues about its author: he was among the "small fry" during Willis's tenure with the USGS, a friend and supporter of Hill, and on friendly terms with either Ellis Shuler or Justina Hill. He also had some familiarity with Georgia plantations.

Although the writer of the colorful memo likely cannot ever be identified with certainty, one possible suspect emerges in the person of T. Wayland Vaughan, born in Texas in 1870. Vaughan, who went on to become an accomplished geologist and oceanographer known to colleagues as both good-humored and strong-willed, worked as an assistant geologist with the USGS from 1894 to 1903. During his initial years with the Survey, Vaughan was an assistant to Robert Hill, with whom he worked in the West Indies and for whom he developed tremendous admiration and appreciation. His fieldwork during this time also brought him to Georgia. Years after his USGS tenure, Vaughan wrote, "Hill was generous with his ideas and I am sure that other speakers will give greater emphasis to them than to his splendid scientific achievement, for they are the traits that have bound his friends to him." For a time, both Vaughan and Hill worked under Willis's supervision; Hill's junior by twenty-two years, Vaughan would have been among the small fry during this time. Later in life, after Hill's death, Vaughan not only corresponded with Ellis Shuler but also was among the colleagues of Hill's with whom Justina Hill corresponded after her father's death. Vaughan's professional correspondence throughout his career was anything but colorful, but by 1947 he had suffered health problems that left his eyesight impaired, potentially accounting for a semilegible scrawl and perhaps an inclination to pen an unusually colorful note.

A second, perhaps better, suspect emerges in the person of Hill's longtime colleague and friend Charles N. Gould. Born in Ohio in 1868, Gould received a master's degree in 1900 and was immediately hired by the University of Oklahoma. Although he was never employed by the USGS, in 1903 the agency awarded him a project to survey the underground water resources of the southern plains, work that continued through 1907. His long and warm association with Hill began before he completed his degree; the two took many field trips together. Gould was among the colleagues who wrote touching tributes to Hill after his death, making the trek to a memorial service held at Round Mountain, Texas. He also corresponded with Hill's daughter Justina after Hill's death. Gould had crossed paths with Willis as well. In October 1902, after Willis had been part of a field trip led by Gould, Willis wrote to his wife, describing his colleague—a man who would later be hailed as the father of Oklahoma geology—as "a young geologist . . . with a long head though short wit." Comparing Gould's earlier letters with the Windy Willis note, Gould's penmanship in earlier years was somewhat tidier, but there are similarities with the loopy scrawl. Gould too was in his later years at the time the note was written; he died in August 1949. On handwriting alone, he emerges as the lead suspect. But whether the note was written by Gould, Vaughan, or somebody else, it provides a telling glimpse of how Willis was viewed by at least some of the geologists who knew him during his early years at the USGS.

When any individual seeks publicity for a good cause, it might be impossible to know to what extent they are driven by more personal motivations. While we do not know the extent to which personal motivations impelled Willis to seek out media attention, at a minimum, it is clear that he never shied away from an opportunity to step onto a stage. The champion of Progressive causes had no difficulty viewing these antics as serving a greater good, which in fact they usually did. Happy and wise is the egotist who devotes his life to a noble cause.

As the great quake debate reached a boiling point, however, forces conspired to pull Bailey Willis and Robert Hill in different directions. Willis, ironically, exited the stage. In late 1926, the Carnegie Institution, which had supported several of Willis's earlier international expeditions, provided support for an ambitious year-long scientific expedition to some of the world's most earthquake-prone countries. Margaret accompanied her husband during the first part of the trip, which brought Willis to Japan at the end of November 1926 and to Bangkok in December. After his wife returned to California, Willis's faithful letters to her chronicle the rest of his travels. By January 1927 he landed in the Philippines; over the following months he visited other countries in the region, including New Zealand and Australia. By July he landed in Athens, Greece, from where he proceeded to the Holy Land. When an earthquake with estimated magnitude 6.3 struck Jericho on July 11, 1927, Willis was in Cairo, Egypt; he proceeded from there to Jerusalem to investigate the earthquake. The world tour came to an end in October 1927.

Looking back at the time line of events, one suspects that forces did not conspire as a simple consequence of fate to shape Willis's trajectory. Willis's correspondence with Arthur Day reveals that Willis proposed the project on "world seismology" to the Carnegie Institution in late 1926, and, by early December, Carnegie approved the grant. Did Willis actively seek to extricate himself from the debate, and from California, just as his own words caused the great quake debate to overheat? Or did Carnegie perhaps nudge him off the stage with a suggestion or offer that was too good to refuse? Whether or not the original impetus for the trip came from Willis, the funding came from Carnegie. It served Carnegie's purposes to have Willis well removed from the California stage, or any public stage for that matter, through most of 1927. There are, furthermore, hints that Willis avoided talking about actual earthquakes on his expedition to study world seismology. In a letter to Margaret dated January 11, 1927, Willis described his meeting with an official in the Philippines, noting that he spoke "even on earthquakes in response to direct questions." The curiously worded sentence makes sense if Margaret understood that Willis had been cautioned to not talk to local officials about earthquake hazard during his worldwide tour to investigate earthquakes.

Outside forces also began to guide Hill's trajectory in the aftermath of Willis's prediction. The invitation to teach at UCLA reached Hill in the summer of 1926, setting him up to speak in the future as not only the father of Texas geology, but also as a member of the local community of earthquake professionals. On December 15 of that same year, Charles Gould, then director of the Oklahoma Geological Survey, informed Hill that the Survey was planning an anniversary commemoration of one of Hill's influential papers, an event to be held not in Texas, but at a private club in Los Angeles. Gould was in Oklahoma, not California, but there were always strong ties between industry geologists in the two regions.

Gould solicited a long list of colleagues, far and near, to write letters of praise celebrating Hill's extensive professional accomplishments. In addition to letters from Hill's close colleagues, some of the fifty-odd letters were from luminaries. The acting director of the USGS, for example, wrote, "He is one of the most able of American geologists, has an enviable national reputation and has contributed greatly to the foundations on which geologic science rests."

News of the honor buoyed Hill's spirits, still flagging following the health problems that had plagued him the previous summer. In January 1927 a dinner was given in Hill's honor, with "compliments heaped on compliments, and honors enough for a horde of men." Daughter Justina, by then a well-respected microbiologist, added a touching and personal testimony to her father. The *Los Angeles Times* published a short article about the event, rich in superlatives. "He has been hailed by many as 'the greatest living geologist,'" the article explained, continuing, "Southern California, which has been the field of Dr. Hill's activities for more than a decade, is regarded by this scientist as the greatest outdoor geological museum in the world." The article concluded by noting that Hill's earlier work had "contributed largely to the selection of the Panama Canal site as the connecting link between the Atlantic and Pacific waterways."

Reverberations from Willis's prediction continued to be felt across the region just as Hill was being hailed by his colleagues in January 1927. If local powers had not set out specifically to groom him to be an especially effective counterpoint to Bailey Willis, they could scarcely have done a better job. Over the months that followed, Hill spoke increasingly publicly about Willis's prediction, while Bailey Willis himself remained safely out of sight. In July 1927, the *Los Angeles Times* described an address Hill had given to the Building Owners and Managers Association of Los Angeles, again describing him as "an eminent geologist, who has spent the last fifteen years studying the geologic conditions of the Southland." The article quoted Hill as saying, "There is not a thread of evidence on which to hang a prophecy of an earthquake in this district." He again struck his usual chords: "Accustomed as I am to the heat of the tropics, the tornadoes of the east, and the Galveston floods, our occasional little earth tremors merely give me a little thrill in the day time or rock me to sounder sleep at night." The article concluded with Hill's quote: "Most of the faults or rifts in the earth's crust in this section are dead faults, caused by disturbances so many millions of years ago that none of us remember the exact dates."

The Building Owners and Managers Association invited Hill to present another address to their members in December 1927. In this speech, a copy of which is preserved among Harry Wood's papers, Hill noted, "Do not get the idea that I underestimate the actual earthquake danger in southern California, or depreciate any steps which are being taken to lessen their effects when they come." This quote did not appear in an article published in the *Los Angeles Times*, which again cast Hill as a foil to Bailey Willis. The headline itself staked out a clear position: "City found safe from temblors." It opened by yet again describing Hill as a geologist "of international repute," who "yesterday made public the results of an exhaustive study of the geology and seismicity of Southern California, declaring that no other section in the United States enjoyed greater freedom from major earthquake perils." In his speech, Hill not only decried the inappropriately alarmist "prophecies of Bailey Willis," but also explained that the scientific result that had fueled the prediction had been revisited. By this time, the US Coast and Geodetic Survey had redone its initial calculations. The revised calculations indicated that not only had the estimated buildup of strain been only about 20 percent of the previous estimate, but the uncertainties were such that it was possible there had been no significant buildup of strain. The article went on to note that another leading geologist, none other than Ralph Arnold, had echoed the conclusion that "there is no real need for earthquake [insurance] coverage in Los Angeles." During Hill's speech, he read a letter from the director of the US Coast and Geodetic Survey that retracted the calculations on which Willis's prediction had in large part been based.

With the revision of the earlier Coast and Geodetic Survey result, made public by the end of 1927, the scales began to tip in the debate. Other forces were at work behind the scenes as well, other fingers working to tip the scales. At the end of the January 1927 *Los Angeles Times* article describing the dinner in Hill's honor, there was a mention in passing: "There is now in course of preparation a new manuscript covering [Hill's work in Southern California], which center around the San Bernardino range of mountains." As of the beginning of 1927, Hill still hoped to complete a book-length monograph summarizing the years of field investigation that he had undertaken under contract with the USGS. After so many frustrations and delays, that book would finally come to fruition the following year, but in a form, and with consequences, that Hill could never have imagined.

CHAPTER 9

THE BOOK

> But, my child, let me give you some further
> advice: Be careful, for writing books is endless,
> and much study wears you out.
> —*Ecclesiastes 12:12*

Following Hill's reassuring speech to the Building Owners and Managers Association of Los Angeles in July 1927, the association commissioned Hill to write a report summarizing his findings. For this work, Hill was to be compensated at a base pay of one hundred dollars per day for work expected to take ten to fifteen days. His daughter Justina visited Los Angeles that summer and helped him to complete the report. Upon delivery of the initial report, association secretary Charles A. Copper (sometimes mistakenly referred to as Cooper) persuaded Hill to expand the brief treatise into a book. The agreement called for publication by the Southern California Academy of Sciences. Hill was introduced to the president of the academy and promised to deliver a worthy scholarly book. Copper made it clear that the book should be completed with all due haste, so that it would be published while public interest remained high.

The association would provide Hill with editorial services, proofreading, publication, and distribution costs. The book would not be the exhaustive monograph Hill had once planned to complete, and the precise details of the financial compensation appear to have been only vaguely spelled out. Murky details aside, Hill accepted the offer and set out once again to write up his work.

Hill had two interests in writing the book: First, he wanted to summarize the extensive geological mapping that he had done, which had never been—and at that point was almost certain to never be—published properly, at least not with US Geological Survey resources. (His 800-plus-page draft appears to have languished permanently in his personal papers.) Second, much as he had been responding to talk of end-times after the 1909 Strait of Messina earthquake, he continued to see a need to provide a prudent scientific counterpoint to what he felt had been irresponsible grandstanding on Willis's part. In this case, he sought to counter not only alarm but also damage to the economy that, in his view, had been wrought by overly alarmist statements.

Like so much else in Hill's life, the eventual publication of his book involved machinations. At the outset, although the arrangements ostensibly called for the Southern California Academy of Sciences to publish the book, as was eventually indicated on the dust jacket, it was an association in name only. An Academy of Sciences imprimatur gave the book a measure of gravitas, but the book would in fact be published and distributed by Copper himself. Copper not only oversaw the publication process, arranging for editing and being the primary point of contact for Hill, he also went on to handle publication and distribution of the book and retained copyright himself.

As Copper had urged, Hill completed the book quickly, eager to see at least a cogent summary of his fieldwork published and reassured that the proffered professional proofreading services would smooth over any rough edges. By the end of 1927, Hill had substantively completed his manuscript. In December of that year, with backing from the Building Owners and Managers Association, he traveled to the Geological Society of America meeting in Cleveland, Ohio, where his presentation focused on a refutation of Willis's prediction. His published abstract outlined his usual arguments, emphasizing that the ominous US Coast and Geodetic Survey result had been revised. "It is to be hoped," his abstract concluded, "that with the new light on the amount and direction of movement Doctor Willis may modify his statements in this connection to quiet the fear which his predictions created . . . and restore confidence in financial circles." The published meeting proceedings then note only that "brief remarks were made by Prof. Bailey Willis." By Hill's later account, Willis had stated that he "was still unconvinced" by the arguments presented.

Extant documents do describe Willis's other activities at the meeting. He was elected president of the society and spoke not about California but, rather, about the Dead Sea Rift, drawing from his observations during his 1927 travels. Willis's talk was described in a number of newspaper articles. If he had been upbraided by Hill's presentation, the meeting as a whole had not gone too badly for him. He segued seamlessly from leadership roles in the Seismological Society of America to equally lofty leadership positions with the Geological Society of America. Scientifically, his interests had already moved on, it seemed, from California to other parts of the world.

From Cleveland, Hill traveled to Washington, DC, where he met with officials from the USGS and the US Coast and Geodetic Survey. He later described "friendly calls" upon the directors of both agencies, during which the subject of California earthquakes was discussed. But he added, "no judgements were expressed by them as to the merits of either side of the controversy." Following his stop in DC, Hill returned to Los Angeles.

On January 19, 1928, the *Los Angeles Times* published a front-page article with the headline "Temblor Talk Tumbled Over" and a subheader, "Scientist shows no major quake peril here," which presented a fairly extensive summary of Hill's points refuting Willis's prediction. The article quoted Hill's statements about seismic hazard in Southern California, including the fact that "historically Southern California has been free of major earthquakes" and the usual mantra that the most severe US shocks had been in Charleston and the Mississippi River valley. Although Hill was in Los Angeles at the time, this article did not appear to generate any consternation on his part. On January 28 he turned in the galley proofs on his book "with an ok as to the subject matter" and left for Texas. On January 31 he wrote to Copper, "You can see now that the prophet should have said that of the making of a book there is no end." The letter continued, "Don't send me anything that you can avoid sending. In doubt make the decision yourself if possible. I will trust your judgement."

While Hill remained in Texas, the publication process indeed moved forward, but the fireworks began even before the book was out. In early February 1928, Hill discovered what he termed "hideous and untruthful publicity" about his conclusions in the *Los Angeles Commercial and Financial Digest*. An article written by Copper himself claimed that during Hill's recent visit to the Cleveland meeting of the Geological Society of America and subsequent stay in Washington, his conclusions had received the "endorsement" of the USGS and US Coast and Geodetic Survey. Also, according to the article, Willis had "completely [abandoned]" his earlier "elaborate arguments," offering instead "only the most perfunctory response" to Hill's rebuttal. Although what Willis actually said in his brief remarks at the society meeting was not recorded, by late 1927 he does appear to have stepped away from his earlier role as active provocateur in the great quake debate.

Hill wrote to Copper immediately to protest what he saw as egregious overstatements in the article. He argued that the USGS, for example, had not weighed in as decisively as the article claimed. Indeed, by all accounts the agency had stopped short of an official position statement. The agency's reaction is documented, however, in a letter sent to Los Angeles bank director and city booster A. L. Lathrop, who had written the USGS director to express concern about the fallout of Willis's prediction. "If," Lathrop had written, "you are in a position to make a statement in this matter, it would be of enormous significance to the welfare and future of this City." The director responded in a letter dated March 31, 1928. After noting the general difficulty of earthquake prediction and the ongoing work by many agencies in California, the director wrote, "It is true that the Coast and Geodetic Survey has recently recomputed the positions of some of its California stations and that the revised positions do not indicate the existence of certain earth strains the preliminary computations had constituted one of the important bases for earthquake expectancy. Mr. Hill, at various times and places, has called attention to the Coast Survey's later findings, and has thus performed a public service." The letter continued, "In so far then as earthquake expectancy in the southern half of California has rested upon evidence of accumulating earth strain derived from the Coast Survey's earlier determinations of the geodetic positions of certain of its stations, the revised determinations are reassuring." In this statement, measured to tolerance levels that would make today's National Aeronautics and Space Administration proud, the letter did not say that Southern California was free from earthquake risk. The letter had, however, provided an effective endorsement of Hill's conclusions; in effect,

Copper's article had, in at least some ways, erred in tone more than substance. In Hill's eyes, however, the publicity was both outrageous and disastrously injurious.

It is an interesting question to consider: why did Copper's article stoke Hill's ire when the earlier *Los Angeles Times* article had not? The articles, while similar in some ways, are also different in several respects. First, the *Times* article stuck more closely to Hill's arguments, with a minimum of editorializing. Second, the *Commercial and Financial Digest* article had been written by Copper himself, an individual whom Hill had once counted as a close personal friend. Last, Copper's article had overstated at least to some degree the reactions of both Willis and government officials. In the mind of a man who never stopped craving validation—who never quite believed he was accepted as a member of the club of elite scientists—these statements irreparably harmed his reputation among his professional colleagues.

With the publication of Copper's article, Hill may have also gotten his first inkling of how his book would be weaponized by some in the business community. Robert Hill, who did not have a pragmatic bone in his 5-foot 4-inch (1.6 m) body, railed immediately and vehemently against the opening salvo of that campaign. The situation would indeed soon go from bad to worse, but not before current events in Los Angeles conspired in tragic fashion to push the great quake debate out of the news entirely. Minutes before midnight on the night of March 12, 1928, the 175-foot-tall (53 m) St. Francis Dam, built across the San Francisquito Canyon north of present-day Santa Clarita, collapsed catastrophically, unleashing a 120-foot-high (36 m) flood wave that traveled into the Santa Clara River valley at a speed of 18 miles (about 30 km) per hour. Five minutes after the collapse, the wave destroyed a concrete power station, killing sixty-four of the sixty-seven workers and their family members who lived nearby and cutting power to much of Los Angeles. Although power was quickly restored, floodwaters spilled into the riverbed, flooding parts of present-day Valencia, later known to Southland teenagers as the home of Magic Mountain amusement park, destroying the two main power lines into Los Angeles. Ultimately, more than four hundred people lost their lives.

Banner headlines topped the front pages of newspapers around the country, describing the tragedy. Famed humorist Will Rogers ventured a wry note in a letter published in a number of newspapers around the country, noting that "Florida has been mighty nice about it. Yesterday they headlined it as an earthquake just through force of habit. But today they have stuck to the truth." The letter, however, ended warmly: "We have great sectional rivalries," Rogers wrote, "but it don't take much to wipe it out when something happens." Indeed, sympathies as well as relief efforts poured into the stricken area.

Within days of the collapse, the Los Angeles Bureau of Power and Light retained three geologists to conduct an investigation into the cause of the disaster—among them, Robert Hill. The team eventually concluded that the dam had first broken at or near a fault line that ran underneath the structure. Increased percolation of water along the fault had, they concluded, undermined the integrity of the foundation to the point that either part of the dam structure blew out or the dam collapsed under its own weight. Other committees were formed to evaluate the cause of the disaster; all but one of the teams agreed substantively with this conclusion. A single team, including two engineers and a geologist, reached a different conclusion from the others. The geologist, Bailey Willis, concluded that a small landslide under the east abutment of the dam had been a small movement along a much larger prehistoric landslide. A small reactivation of that slide, Willis concluded, not water percolation or, as some others suggested, faulty engineering, had caused the dam to fail. Many years later, geological engineer J. David Rogers reconsidered the dam collapse using computer simulations and other modern analysis techniques. In a book published in 1995, Rogers laid out the modern evidence in support of his conclusion: Willis alone among the geologists of his day had been right; the failure had been caused by a reactivated movement of an ancient landslide.

The St. Francis Dam collapse and subsequent investigations played out just as the great quake debate reached a crescendo, likely serving to defuse a situation that would otherwise have become even more visible and contentious. Outside of California, the dam disaster garnered more headlines than the great quake debate ever did. And for both Hill and Willis personally, focus on an actual tragedy outweighed, at least temporarily, concern about hypothetical future earthquake disasters.

The book publication process, however, moved forward behind the scenes. The machinations that played out with that process are difficult to sort out, and Hill's own papers do not agree with the story in the biography written by Nancy Alexander. By Hill's later account, he called the printers on March 19, 1928, a week after the St. Francis Dam collapse, to ask about the page proofs, which were delivered to his Texas hotel on March 24, just as he left Texas to return to California to conduct his investigation. In one of the documents Hill wrote after the book was published, he wrote, "Since no haste . . . was indicated or the desire for early return of the proofs in any manner requested or intimated, they were in my hands, being carefully read and corrected for three weeks, waiting to be called for." The assertion, on the heels of earlier indications that both Copper and Hill wanted to see the book published in all due haste, does not ring true. In any case, on April 16 he arrived back in Texas and found six copies of the printed book waiting on his desk. "These," he later wrote, "had been printed without the faintest attempt to procure the final page proofs from me with my final corrections and approval."

The books that greeted Hill upon his return sported bright red jackets. The front of the jackets were emblazoned with not only the title, *Southern California Geology and Los Angeles Earthquakes*, but also the equivalent of bold headlines: "This book completely refutes the predictions of Professor Bailey Willis that Los Angeles is about to be destroyed by earthquakes. It proves that this area is not only free from a probability of severe seismic disturbances but has the least to fear from 'Acts of God' of any city under the American flag." The blurb continued, "Here are first given by a great scientist in popular language an explanation of the natural causes of our beautiful scenery and charming climate. Southern California is a great out-door geologic museum. The layman is offered a key that will add enormously to his enjoyment of this unique playground; the student of geology or physical geography will find this an indispensable guide." On the back, the jacket began with a brief and glowing biography titled "Dr. Hill has had one of the most distinguished careers in American science." It went on to include excerpts of tributes to Hill provided by a dozen leading experts in the field—pulled from the many glowing tributes that had been collected, conveniently, in advance of the January 1927 dinner in Hill's honor.

Inside the cover, the bulk of the 230-page, seven-part volume focused on the geography and geology of Southern California, including general remarks about faults, descriptions of many individual faults, and the geological history of Southern California. While some of the material has been superseded by later work, parts 3 through 6 laid out much of what was known at the time about the region's major faults and physiographic provinces. These chapters appear to have been a brief summary of Hill's earlier comprehensive book draft. Parts of part 2, "Generalizations concerning earthquakes," include some by-now outdated ideas but also provided a reasonable summary of the state of knowledge at the time. In writing this material, Hill, whose scientific work had never involved earthquakes per se, drew heavily from a recent book by a Harvard professor. This part of Hill's book echoed words that Ralph Arnold and others had written, in articles and their own speeches, about earthquakes.

Among the poignant notes in the great quake debate is this: from the time Hill's book was published, its substantive content—the material he had yearned for years to write, drawing on years of fieldwork—was effectively irrelevant. Although the book presented a first-ever synoptic overview of many of the region's faults and large-scale physiographic provinces, it has been cited by only a small handful of subsequent scholarly publications, only a few of which cite the book as a seminal geological reference. Although modern scientists can generally be guilty of overlooking early seminal references, the small number of citations of the book contrasts to the recognition of many of Hill's other publications—for example, his monograph on the Black and Grand Prairies in Texas, which has been cited more than two hundred times.

As soon as the book went on sale in April 1928—in fact, even earlier, when advance copies had been circulated by Copper—the fact that Hill's book provided a broad overview of the geological landscape in Southern California was beside the point. Interest focused immediately not on what the book had to say about Southern California geology, but on the words on Los Angeles earthquakes presented in the brief part 1, "Southern California Attacked," some of the statements in part 2, the brief concluding section, and the cover itself. As soon as he had seen the bold words on the dust jackets of advance copies, Hill protested "against the quack doctor publicity . . . I protested against the betrayal of confidence in publishing the private letters on the red covers . . . I protested against the false assertions about Dr. Willis and myself." Hill requested that the book not be distributed until his concerns were addressed. These protests, however, were to no avail: Copper and his associates made sure that distribution of the book moved forward. Inside the book, a four-page folio reproduced a pair of letters, the March 23, 1928, letter from A. L. Lathrop to the director of the USGS and the director's excruciatingly measured reply concluding that "the revised determinations are reassuring."

As for the actual contents of the book, they can be considered with the benefit of hindsight. Beyond the garish cover, the book opened with a brief foreword, a two-page discussion of the book's purpose and content, including a penultimate paragraph: "Since I believe that the lessons on geologic history teach us that each relief-shock is a step towards some adjustment or equilibrium and that there is no reason for expecting shocks of increasing seismicity in the future, it will be necessary to present briefly some of the facts of the more recent geologic history." Scientists now know that the first part of the sentence is right, in the sense that earthquakes release stress, but also wrong. Because of ongoing plate tectonics forces, of which Hill had no inkling, the earth will not settle into equilibrium anytime soon. We also now know that recent earthquake activity, such as the moderate shocks that Southern California had experienced in the 1910s and early 1920s, actually weakly increases the odds of a future large earthquake, because earthquakes can trigger other earthquakes. In retrospect, in any case, Hill's conclusion that "there is no reason for expecting shocks of increasing seismicity in the future" was and is a reasonable statement.

Part 1 of the book focused squarely on Willis's prediction, describing it and its effect on local insurance rates, emphasizing that the US Coast and Geodetic Survey result provided the "cornerstone" of the prediction. "If this allegation . . . should prove to be erroneous, then the remainder of the edifice will crumble." Part 1 included the observation, "That southern California has been a land of moderate seismic disturbances has long been known." In other words, Hill conceded that earthquakes were not unknown in the region, but he went on to make the case, once again, that they did not pose a serious danger. Throughout part 1 he struck many of the same notes he had in earlier speeches: "These phenomena have been considered of such secondary importance as dangers to life and property, that the public, including men of all ranks and professions, has been content to risk its lives and money here, and, notwithstanding prophecies of disaster, will doubtless continue to do so with a perfect sense of security."

Hill also included many of his familiar refrains in part 2. He decried the inflated portrayal of damage from past earthquakes felt in the Los Angeles area. He wrote:

> There is much evidence to show that the severity of tremors in 1769, 1812, and 1857 in southern California were greatly overestimated. The falling of the tower of the old adobe mission church at San Juan Capistrano in 1812 (29 persons having been killed thereby) and of adobe houses at Yuma and Tejon in 1857, are the outstanding earthquake disasters in southern California. In each instance, the buildings of adobe mud were inferior in construction, without proper reinforcement, if any, and of crushing and tensile strengths so low that one wonders how they sustained the weight of their own walls.

He pointed to the lack of serious damage to the San Gabriel Mission since its construction in 1771 as further evidence that the greater Los Angeles region had never been hit by damaging shaking through documented historical times.

After reviewing the effects of past earthquakes, Hill went on to say, "There is no proof that a serious earthquake has ever taken place in Los Angeles. In all the history of Los Angeles, there is no record of a good

building having been thrown down." As jarring as this and other statements are to a modern ear, they were not actually wrong at the time. The
claim that Hill went on to make, that Los Angeles itself was a region of
"mild seismicity," was wrong, but still not as ludicrous as it now seems.
Using the Global Positioning System (GPS) and other techniques, modern
science has given us estimates of the rates of movement along all of the
world's active plate boundary zones; these rates control the level of earthquake hazard in a region. These results have allowed us to measure not
only the long-term rate of plate motion along the San Andreas Fault, but
also the rate at which strain is building on secondary fault systems in the
plate boundary zone. We now know that the greater Los Angeles metropolitan area is undergoing broad compressional forces associated with a
bend that leaves part of the San Andreas Fault not perfectly aligned with
the direction of plate motion. These forces give rise to active faulting that
mostly serves to push local mountain ranges, including the Santa Monica
Mountains as well as the San Gabriels, upward. (Ironically, in the greater
Los Angeles area, most active faulting is indeed vertical.) The rate of
earthquakes on Los Angeles–area faults does not match that along the
world's most active plate boundaries, including the South Pacific, Japan,
and coastal South America. Los Angeles is further, as Hill argued, some
40 miles (64 km) removed from the San Andreas Fault. "Mild seismicity"
went too far, but "moderate" would have not been unreasonable.

Hill's arguments erred in another way that was understandable at the
time: neither he nor any earthquake professional of the day understood
what scientists know today—namely, that in a probabilistic sense, moderately severe earthquakes such as the 1925 Santa Barbara earthquake
can be even more important for hazard than much larger earthquakes
like those in 1857 or 1906, because very large earthquakes strike so
infrequently. That is, while the effects of a great earthquake on the San

Andreas Fault might be portentous, in any one lifetime most Californians are statistically more likely to experience a moderately large earthquake such as the 1994 Northridge temblor. Californians might fear the Big One, but statistically it's the Pretty Big One they should worry about.

Part 2 of Hill's book included a detailed refutation of Willis's prediction. Hill began by explaining the reasons why earthquakes cannot be predicted, noting reasonably, "Such meager data as we possess concerning California seismicity has been closely scanned in hopes of ascertaining some law whereby future earthquakes may be predicted, and several attempts to discover laws of periodicity have been made. But alas, the seismicity has been too slight and the records too incomplete for coming to satisfactory conclusions." These statements were reasonable at the time; in fact they remain eminently reasonable today. He went on to explain that the "fundamental data" behind Willis's prediction had been proven to be erroneous. Whether or not he actually wrote the bold statement on the book's cover—"This book completely refutes the predictions of Professor Bailey Willis that Los Angeles is about to be destroyed by earthquakes"— the end of part 2 echoes words he had said many times before: "Thus the foundations upon which were built the whole structure of Dr. Willis's predictions of disaster in southern California have crumbled, bringing all of the superstructure with it."

Although most of Hill's words throughout his book present the same arguments he is known to have presented elsewhere, a handful of words are conspicuously missing. The disclaimer he had included in earlier speeches—"do not get the idea that I underestimate the actual earthquake danger in Southern California or depreciate any steps which are being taken to lessen their effects when they come"—is nowhere to be found.

The publication of Hill's book appeared to firmly establish Hill as the scientist who embodied one side of the great quake debate, the expert geologist who had not only decried a specific prediction but also proclaimed Southern California to be free from serious earthquake hazard. When the earth itself proved such statements spectacularly wrong just a few years later, Hill's casting as villain—if not laughingstock—appeared to be sealed. Hill's own later writings attempt to paint a different picture. According to extensive correspondence, from the time he first saw the initial public relations campaign and then the early copies of his book, he made his feelings known in no uncertain terms. And Robert T. Hill had never been one to take a slight, real or imagined, lying down. After he received his copies in April 1928, he not only demanded that distribution be halted but launched a vehement protest campaign. He was especially aghast that the book would be published under the (supposed) auspices of the Southern California Academy of Sciences, to whom he had promised a book worthy of their imprimatur. For its part, the academy appears to have been caught in the cross fire, an innocent bystander in this chapter of the great quake debate.

An advance copy of Hill's book landed in the hands of renowned geographer William Morris Davis, who had been asked to write a review of the book. Hill knew and respected Davis, who had been one of his supporters and counselors when Hill had found himself in a pitched battle with Edwin Dumble back in Texas. Davis, who had also been an early admirer of Bailey Willis's scientific contributions, had been taken aback by parts of the book. Hill, in turn, was taken aback by Davis's review. In Hill's eyes, by writing the review, Davis "[had] butted into a situation which does not concern him." Also in Hill's eyes, six decades after the end of the Civil War, old divisions were once again at play: "Why God Almighty put the missionary spirit into the Yankee which causes him to meddle with everybody's business but his own is one of the mysteries of the universe which I do not understand." Hill added, "I am very fond of Professor Davis, but he simply delights in getting out of hand."

What, one asks naturally in retrospect, did William Davis actually say about the book? His one-page review was not entirely negative. "The predictions made several years ago by Bailey Willis," the review began, "to the effect that a severe shock is likely to occur in southern California in from three to ten years, are now contradicted in a book by R. T. Hill, widely known for his excellent geological work, especially in Texas." The review quickly segued, however, to a different tone. Davis wrote that the book "has the style of a legal brief, prepared to defend the interests of a client rather than that of a scientific monograph prepared to set forth impartially all the pros and cons of a highly theoretical and very debatable problem." Davis further mentioned the statements on the initial version of the front cover, declaring them to be so "glaringly false" that they were to be later replaced by milder statements on a later version of the cover. He was under the impression, likely from what Hill himself had told him, that an updated version of the book would be forthcoming, but this appears to have been only wishful thinking on Hill's part. The original red dust jacket is not easy to find, but copies do survive; there is no indication that a revised cover was ever published, although it is possible that later copies were distributed without any dust jacket. In his review, Davis did mention an errata sheet he had received, which included this note: "In justice to the author it is proper to state that this book was printed without awaiting his final corrections on the last galley proofs and page proofs." Davis, however, went on to write, "One who knows Hill's other publications must wonder how far his failure to see the final page proof is responsible for the actual form of this book." It is small wonder the review left Hill feeling stung.

To the surprise of nobody who has read this far, a later review of the book published in the *Los Angeles Times*, written by Ralph Arnold, struck a different tone. "This little book," Arnold wrote, "published by the Southern California Academy of Sciences, should be read by everyone who is at all interested in Southern California. In it Hill at once has given us a comprehensive and entertaining word picture of the topography and geology of our wonderful Southland as well as facts regarding the earthquake situation." Although this statement continued "which should enable even the layman to realize that this is one of the real problems confronting our citizenry," he went on to emphasize one of his usual refrains: "If we use even ordinary efficiency in construction, with thought as to the location and natural environment of our dams, bridges and buildings, then I feel justified, as an engineer and scientist, in saying that we as citizens of Los Angeles and vicinity have little to fear from any future earthquake."

The forces at work to bring Hill's book to fruition had been delayed but not derailed by the St. Francis Dam collapse. In April 1928, Hill's book was offered for sale for the then not-insubstantial price of five dollars, and yet more publicity followed. On April 16, a front-page article in the *Los Angeles Times*, headlined "Expert Shoos Quake 'Bogey': Southland Safe, Says Hill in Geology Book," opened by parroting with modification the blurb on the front cover. Whereas the book jacket proclaimed that "this book completely refutes the predictions of Professor Bailey Willis that Los Angeles is about to be destroyed by earthquakes," the article claimed that "this book refutes the predictions of Dr. Bailey Willis, a prominent geologist, that Los Angeles is endangered by earthquakes." Hill had been upset enough about the earlier publicity; this article upped the ante. Although the difference in wording is subtle, the claim on the book jacket was that the book refuted a short-term prediction—that is, that Los Angeles was *about to be destroyed* by earthquakes. The article took this claim one subtle but critical step further, suggesting that Hill had refuted the notion that Los Angeles was *endangered by* earthquakes. The article moreover went on to quote the second sentence from the book jacket, again not entirely faithfully, stating that "this area is not only free from a probability of seismic disturbances, but has the least to fear from 'acts of God' of any city" in the country. Although this sounds much like the sentence on the cover, the article omitted a single word, replacing "severe seismic disturbances" with, simply, "seismic disturbances." In other words, Hill's conclusions were summarized not as saying there was a low probability of a large earthquake, with no reason to think that a large earthquake was imminent, but, rather, that as the headline proclaimed, Los Angeles was safe from earthquakes, period.

The April *Los Angeles Times* article went on to discuss Willis's prediction and to say that Hill had noted in his book that the San Andreas Fault never comes closer than within 60 miles (97 km) to Los Angeles. Here again the article missed the mark in a small but important way. Hill had written that "certainly the longer distance of Los Angeles from the San Andreas, San Jacinto, and Santa Ynez rifts greatly lessens our dangers from them. I believe that the statistics prove incontestably that the San Jacinto fault line, the nearest point of which is sixty miles from Los Angeles, is the most active fault line in Southern California." Thus Hill noted that the *San Jacinto Fault*, not the more widely known and feared San Andreas Fault, was no closer than 60 miles (97 km) from Los Angeles. The modern scientist cannot fault Hill's reasoning in the argument that he did make regarding the danger from the San Jacinto Fault. By 1928, seismologists knew from firsthand observations of the 1906 San Francisco earthquake that the strongest shaking from an earthquake on the San Andreas Fault, or other faults like it in California, is largely concentrated along a narrow ribbon in proximity to the surface trace of the fault. Thus, proximity to the fault makes all the difference for hazard. Hill was moreover correct in identifying the San Jacinto as the most active fault in California, in terms of its production of small to moderately large shocks. He could not have known what was established in later years by decades of scientific investigations: the San Andreas Fault has produced relatively few small to moderate earthquakes over historical and recent times—fewer than the San Jacinto—but over the long term is more likely to generate large earthquakes, with magnitudes upward of 7.0, than the San Jacinto. Overall, the *Times* article had, in several small but impactful ways, misconstrued Hill's reasonable and factually accurate sentences, creating inaccurate and unjustifiably reassuring statements.

The *Times* article went on to note that "numerous local organizations, prominent among them the Los Angeles Building Owners' and Managers' Association," had undertaken efforts to roll back recent insurance rate hikes. The association hired a local team of attorneys to ask the California State Corporation Commission for an expression of future policies for earthquake insurance. In response, the commissioner declared in an article in the *Los Angeles Times* that "the high rates are traceable to predictions by Prof. Bailey Willis of Stanford University that a major earthquake will occur in portions of the State within ten years." He went on to note that other scientists had refuted the prediction, and, in light of the "divergence of opinion by those who have made a study of the situation," the issue "must be resolved in favor of the normal." Furthermore, the commissioner claimed there was "no anxiety among financiers, architects, engineers, and contractors over a possible disastrous earthquake in California."

"It seems hard," the commissioner said, "to reconcile the fact that on one hand California is being penalized by earthquake insurance companies inflicting exorbitant rates, while on the other hand eastern insurance companies and other financial houses are continually sending millions to this State on loan for improved real estate, scarcely ever insisting on earthquake insurance." In light of all of this, the state corporation commissioner announced that earthquake insurance would no longer be a prerequisite to the issuance of permits for the sale of building bonds.

The Building Owners and Managers Association thus got the desired return on its investment in Hill and his little book: commercial earthquake insurance rates began to drop. For his part, in characteristic fashion, Robert Hill continued to fume. By Nancy Alexander's account, he hired detectives, who learned that the book had already been ordered to press at the time he received page proofs to review. Hill moreover suspected that the proofs he received differed from those used by the editor. Hill's papers suggest, however, that he asserted but did not prove that these things were true. In any case, Copper had, Hill concluded with increasing conviction, deceived him willfully. On March 7, 1928, he wrote in his diary, "Coming around to a realization of greater astral forces."

Hill's grievance was twofold: first, that his book had been altered without his knowledge, but second that he had only ever been compensated for the report that he completed in the summer of 1927, not for the considerable further work he had done on the book. He pressed his case in court, seeking $106,500 plus attorney fees in compensation, the lion's share for "damage to reputation." It is not clear how the case was resolved; Nancy Alexander concludes that Hill's financial circumstances remained modest, suggesting that any compensation he received had been small. Certainly there is no indication that he received anywhere near the six-figure damages he sought. In December 1928, Hill did send a letter to Cornell University regarding student loans still outstanding after a half century, noting that "a combination of circumstances, not to mention the gnawing conscience, which includes the receipt of a good-sized professional fee . . . has aroused in me an irresistible desire to action, as attested by the enclosed cheque." It seems, then, that he did receive some amount of financial compensation toward the end of 1928.

Financial considerations aside, in terms of other consequences the words that had been printed in his book could not be taken back. Words on a printed page are not only indelible; at a time before the Internet and even television, they had weight. The book had been published and distributed and the publicity had been generated, all to the desired effect of the forces that set the book into motion in the first place, but in Hill's eyes with ruinous harm to his reputation. As he had ignored colleagues' advice in the aftermath of the Dumble debacle so many years before, this time Hill ignored Ralph Arnold's suggestion in an August 29, 1928, letter that he should "let the matter rest." Arnold persevered in his role as liaison, even conflict mediator, during these months. In June 1928, he wrote to C. A. Copper, by that time Hill's primary nemesis, regarding specific statements that Copper had made, that "[Hill] seemed to think that such conclusions attributed to him really hurt him, and he is very desirous that you do not make any more quotations of his work, except verbatim statements from his book." Two months later, on August 21, one day after he had met with prominent city booster Henry Robinson, Arnold's diary indicates that he had a lunch meeting with Hill, Arthur Day, and Harry Wood. The student of history is left to wonder what these four men talked about.

Ralph Arnold might have been the dictionary definition of politic, but successful mediation of a Robert T. Hill conflict was beyond even his powers. The fact that the book presented arguments that he had indeed made was, in Hill's mind, besides the point. In his view, the changes made by Copper and the tenor of the words on the cover, compounded by the subsequent publicity, had been sufficient to corrupt the book and damage irrevocably his professional reputation among the academic colleagues whose approbation he never stopped craving.

Again the reader is left to wonder where exactly the truth lay. To what extent was the book changed? We have a small mountain of evidence to answer this question, including an eighty-four-page narrative that Hill eventually wrote to press his legal case against Copper. In this and other writings, Hill went to great lengths to describe the changes that had actually been made and a long list of typos that were not caught in the editing process. Hill claimed, for example, that he had added a comment on the final page proofs acknowledging Willis's high standing as a scientist, but the remarks had not been incorporated in the final book. He also claimed that a statement by Harry Wood, to the effect that earthquakes cannot be predicted, was published as written but not highlighted as prominently as Hill had indicated. Hill also emphasized that Copper had changed his references to "this book" to "this report," with the intent on reneging on the promise of additional financial compensation.

Nowhere in his lengthy manifesto, however, did Hill claim that the book had ever included the key statement that he had made in his speeches: "Do not get the idea that I underestimate the actual earthquake danger in Southern California or depreciate any steps which are being taken to lessen their effects when they come." While the opening chapters of the book included many of Hill's other words from his speeches more or less verbatim, it appears that, by either intent or inadvertent omission, these words were left out by Hill himself, not cut out by someone else. Hill moreover never claimed to have not written a comment about the Newport-Inglewood Fault specifically: "Judging from past history, it cannot be said that there is any great menace from the Inglewood line. Its releases from strain have consisted of small vertical movements, sometimes frightening, sometimes slightly damaging, but at no time seriously damaging."

He did claim that some words in the book had been added: "I don't believe that I ever said in the original text that 'the accumulated weight of data substantiates beyond a doubt my deduction that Los Angeles is in no danger of great earthquake disaster.'" Yet this exact statement appears in a draft of the earlier report that Hill had prepared for the Building Owners and Managers Association. In fact, the page from the report draft in his own papers matches word for word the summary of conclusions that appeared at the beginning of part 7 of the book. In a letter to the *Los Angeles Times*, which, if sent, was never published, Hill went so far as to say, "If it should be technically proved that I have made such a statement as the one you cite [i.e., the 'no danger' quote above], I hereby withdraw it, for it is scientifically incorrect." Nowhere, moreover, did Hill claim to have not substantively written part 1 of the book, described by William Davis as a "legal brief." In his battle for justice, Hill railed fairly against the splashy bold cover with its bold words, the equivalent of grabby newspaper headlines. Without question, media articles went on to misquote his words in subtle but important ways. In substance, however, even the words that Hill wrote in his own defense suggest that, overwhelmingly, he had been responsible for the content and tone of the book.

Copper refuted the allegations, backed up by an affidavit from the individual whom Copper had hired to edit the book. The affidavit stated that the "book as finally printed is essentially what was originally submitted" and that "Dr. Hill's claims of alteration in any sense is unwarranted; changes being only legitimate and legally allowable editorial emendations." Copper described Hill's assertions to the contrary as "technical hair-splitting." On one count Copper skirted the issue because he had no defense to offer: Hill had never okayed the blurb on the front jacket nor the glowing testimonials on the back cover. Apart from any of the words inside the book, both of these stoked Hill's ire for the same deeply personal reasons as had the earlier *Commercial and Financial Digest* article.

The weight of the evidence was not on his side, but once again, Hill was unable to back away from an unwinnable fight. Yet again, the man with an uncanny genius to see the geological landscape was utterly myopic when it came to appreciating his own measure of responsibility for a situation gone awry. Without question, the local business community had set him up to play his part on a public stage, but in the end, although he was constitutionally unable to acknowledge it, when he walked onto that stage, he played himself.

The fallout, real and perceived, weighed heavily on Hill's mind through at least mid-1929 and appears to have sent him into an emotional tailspin; the episode also sparked a period of deep retrospection. In May 1929 Hill returned to Texas, where he had spent "the best years of [his] life, amidst strenuous and primitive frontier surroundings." The trip to Texas was Ellis Shuler's doing. Possibly concerned about his friend's physical and emotional health, Shuler had arranged for Hill to be awarded an honorary doctorate at Southern Methodist University. Hill took the opportunity to visit Comanche as well. In a letter to his daughter Justina, he wrote of "hills and prairies [that] were of different contour from those to which I had been accustomed, and . . . excited my attention. . . . These things," he wrote, "were all so long ago that the memories of them had grown into a haze." He continued, "The bitter later experiences with the Washington set who dumped upon me all of the venomous hatred of everything Southern and Texan, had almost obliterated . . . the . . . beautiful recollection of my youthful days. Experience with the Gannetts, the Covilles, the Willises, the Charles Abiathar Whites, and their like, embittered the mind." Even having recently protested vehemently that his book had not, as he had intended, included words of praise for Bailey Willis's scientific accomplishments, he couldn't hide his deep-seated personal antipathy toward the man. In the words of a loving daughter, "Part of my father's charm was his complete inconsistency." Hill's feelings toward Willis were not, however, entirely inconsistent, differentiating between regard for (at least some of) Willis's scientific contributions versus his feelings about Willis as an individual.

In Hill's return to Comanche, however, the former frontier boy found redemption. "A miracle has happened," he wrote, "such as I never expected could take place. After a week of having been superbly entertained at the University at Austin, a couple of fine young modern day geologists—Messrs Plummer and Adkins—who have known and appreciated my earlier work, placed me in a splendid Buick car and drove me . . . back to Comanche in order to view with me the sites and scenes of my boyhood days." It was, he concluded, a "glorious day at old Comanche." The letter, which went on to relay reminiscences of earlier days, grew long, eventually filling up five type-written pages. By the time the letter was finished, he had continued on from Comanche to Fort Worth. "Saturday," he wrote, "I return to Dallas and my host of friends here who love me. . . . On Tuesday, June 4, I will receive an honorary degree from the University here, and I know that I will sink through the chair, for my friend Shuler has been digging up everything nice that he could considering my life and work in Texas."

Hill did return to Los Angeles following the visit, but the publication of his book and its subsequent fallout marked the beginning of the end of the California chapter of his life. The episode also marked the beginning of the end of Hill's career as a geologist. While his darkest clouds of despondency eventually lifted, the introspection sparked by the episode continued, leading to increasing ruminations about his years on the frontier and interest in penning his memoirs. He again reached out to his brother Jesse to fill in some of the blanks in his own mind regarding both their family and the events of Hill's earliest childhood, which he himself barely remembered. He got as far as writing a few brief chapters recounting his early life and career, but the project stalled out, along with the remaining shreds of his attachment to California. His sprits were further dampened by eye surgery in late 1929 that left him fearful about the possibility of losing his eyesight. Although he did recover, his vision would continue to deteriorate through his later years. Stung once again by perceived persecution, this time by the Los Angeles business community that had once so warmly embraced him, his faltering gaze drifted increasingly away from California, back to his beloved Texas.

CHAPTER 10

RETRENCHMENT

> He will win who knows when to fight
> and when not to fight.
> —*Sun Tzu*

Following the publication of Hill's book, the great quake debate retreated into the Southern California woodwork more quickly than it retreated from Hill's own mind. Concern among earthquake professionals did not, of course, disappear; rather, it returned underground. The efforts of city boosters also continued behind the scenes to undo the burdensome consequences of Willis's prediction. The Los Angeles Building Owners and Managers Association continued to press its case on the matter of local earthquake insurance rates and policies. In May 1929, even as one downtown building owner received a $4,000 rebate in insurance premiums, the association urged building owners to conduct surveys of their masonry buildings "with a view of reducing earthquake insurance rates." Insurers had shown a willingness to reduce insurance rates if owners could demonstrate adequate "earthquake resistance qualities of their structures." As the earthquake bogey faded from the public eye, life in Southern California, including business practices, appeared to return to normal.

Earthquakes did continue to pop off around the Southland after the Santa Barbara earthquake; they might have been downplayed by local media, but behind the scenes they did not go unnoticed. On New Year's Day in 1927, an earthquake of estimated magnitude 5.8 had caused heavy damage in Calexico and Mexicali, south of the Salton Sea. As Arthur Day wrote to Bailey Willis two months later, "the recent earthquake at Calexico, which was locally quite severe, has found some reaction in Los Angeles, though it was scarcely felt there." Although Day did not elaborate on the nature of the reaction, an article published in the *Calexico Chronicle* several days after the earthquake provides clues. Under a headline "Damage here may exceed $500,000," the article described the potentially substantial insurance liabilities from this relatively modest earthquake. An article in the *New York Times* cited an even higher estimate of losses, $2.5 million. Even the lower of these numbers was high enough to stoke concerns among those members of the business community who remained concerned about insurance liabilities.

Other significant earthquakes occurred in the Southland during the late 1920s. A large earthquake, now estimated to have been around magnitude 7, struck the Lompoc region on November 4, 1927, but to an even greater extent than the 1925 Santa Barbara earthquake, the Lompoc temblor generated only modest rumblings, literally as well as figuratively, in the Los Angeles region. Closer to home, the Whittier area, some 12 miles (20 km) east-southeast of downtown Los Angeles, was hit by a moderate quake on July 8, 1929. The 1929 earthquake did cause local damage, which seismologists including Charles Richter and Harry Wood set out to survey. But, in both magnitude and effects, the temblor was similar to the 1920 Inglewood earthquake, with a tight concentration of local damage and few serious effects throughout the central Los Angeles region. City boosters were once again quick to downplay the damage that had occurred. The day after the 1929 quake, the *Los Angeles Times* published an article toward the bottom-left corner of the front page headlined "Slight Temblor is felt." The article opened, "A temblor, apparently caused by a slip in one of the minor faults in the Puente Hills district, shook parts of Southern California yesterday at 8:46 a.m., causing some damage at East Whittier and a few other points." Harry Wood wrote privately to the *Times* to protest having been quoted in an article when he had never spoken to a reporter from the paper. But neither he nor anyone else among the recently upbraided community of earthquake professionals in California jumped on this earthquake as a teachable moment to press their cause. Outside of California, the "slight temblor" grabbed the headlines in a bigger way. A banner headline in New Jersey on the front page of the *Trenton Evening News*, for example, read: "LOS ANGELES AND VICINITY IS VIS-ITED BY EARTHQUAKE," with a subhead "4 HURT, NO ONE KILLED. Long Beach Newspaper Plant Rocked Fifteen Seconds."

Although the 1929 earthquake (just barely) grabbed local headlines for a short time, and all of the late 1920s earthquakes provided further impetus to insurance companies to get a handle on the earthquake problem, Los Angeles itself took the temblor in stride. Life was good. Buoyed by both the discovery of new oil fields and deepening of wells in existing fields, oil production in the Los Angeles basin continued the steep climb that began in the early 1920s, nearing 300 million barrels per year by the end of the decade.

Ironically, there is evidence that the deepening of a well in the Santa Fe Springs oil field actually caused the 1929 Whittier earthquake, and possibly other small events in the region, in the early decades of the twentieth century. By that time, even some outside of the scientific community had begun to wonder about a possible association between oil production and local earthquakes. As early as 1902, an article in a San Francisco newspaper laid out cogent thoughts regarding a devasting earthquake on the shores of the Caspian Sea in present-day Azerbaijan, noting that the quake had struck near the Baku oil district, "the most productive field in the world." The article went on to note that earthquakes had not occurred historically in the region and, further, that "the vacuum created by the draining of . . . oil would remain, and a shrinkage of the unsupported crust of the earth would naturally follow sooner or later." Several decades later, in September 1930, the *Los Angeles Times* published a letter from a reader, claiming that a recent felt earthquake had been caused by oil production. This letter was swiftly countered by a letter published on September 10, signed only "R. R.": "These wiseacres who have that phobia that the earthquakes out here are due to the development of oil wells in this vicinity seem short on the knowledge of history." The letter went on to say, "What caused the earthquake of 1812 in California, when there are no oil wells, and why hasn't Pennsylvania had earthquakes since it began its oil-well development over 50 years ago? If they would study geology a little these who know it all, perhaps they would become enlightened as regards the causes of quakes. One of the worst of them occurred in Charleston some forty years ago. What oil wells caused that, eh?"

In effect, the unknown "R. R.," who clearly knew more than a little about both the oil industry and earthquakes, laid out compelling arguments that all earthquakes are not caused by oil production and that oil production does not always cause earthquakes. None of the cited evidence, however, argued against the possibility that oil production might have caused some of California's earthquakes. The idea of an association, not unfairly, remained out there among the general public. In an undated manuscript, apparently from the 1920s, titled "An explanation of the

California earthquakes," Ralph Arnold stated "Erroneous beliefs: Removal of gas and oil from sands. Not so, for sands occupy same space as before."

As much later events demonstrated, oil-bearing rock—sands, in the industry parlance—do not occupy the same space after oil and gas have been extracted. Following rapid discovery and exploitation of the Wilmington oil field near Long Beach, uncompensated withdrawal of oil caused the surrounding ground to subside at such a dramatic rate that the US Navy called on the oil industry to address the problem. And as much later research demonstrated, significant oil production, from deep formations in particular, can plausibly induce earthquakes by perturbing the stress at the depth at which earthquakes occur.

Through the 1920s and well beyond, however, a possible association between oil production and earthquakes was generally dismissed by scientists and, to a perhaps curious extent, rarely even discussed. If scientists had suspicions about an association between oil production and earthquakes, discretion remained the better part of valor. In retrospect one wonders what was known or suspected. In 1931, as the Pasadena Seismology Laboratory began to record earthquakes on its newly installed instruments, it began to locate small local shocks with increasing precision. In May of that year, Harry Wood wrote to a geologist with Union Oil to present the lab's latest findings at an upcoming meeting of the Branner Club, an active local association of industry geologists. Wood wrote, "I recognize that the information accumulated is of interest, and . . . that the Branner Club is a very suitable organization before which to present it, since on account of its nature it is not very suitable for a really public meeting." Had Wood and his colleagues begun to realize that small earthquakes were occurring preferentially near active oil fields? Or perhaps Wood wanted only to avoid publicizing the extent to which small earthquakes were occurring commonly within the Los Angeles basin.

But then there is the letter, dated July 17, 1933, that Wayne Loel, a consulting geologist and longtime colleague of Ralph Arnold, wrote to the head of the California State Corporation Commission, commenting on the prospects for finding oil in the strata of the Pico Anticline, north of Los Angeles near Newhall. The brief letter concluded, "It is my belief that the source of the oil lies in rocks of Eocene age and if such rocks are encountered in drilling the hole, I should recommend that further drilling cease." The Eocene epoch, a division of the geologic timescale, lasted from 56 million to 33.9 million years ago; it was followed by the Oligocene, which extended to about 28 million years ago, and then by the Miocene. That the Los Angeles basin became a major oil-producing region has everything to do with events during the Miocene. During the early part of this period, when the climate remained relatively warm, much of the present-day coastal region in Southern California was a shallow marine environment in which sediments accumulated, eventually to be compressed into sandstones. Also during the Miocene, large-scale place tectonics forces reworked Southern California extensively, creating the Los Angeles and other basins. These forces warped the layers of marine sediments, creating faults and the fault-controlled traps in which oil collected, generating the bulk of California's present-day oil reserves. Although Eocene strata in Southern California can also bear oil, such strata are older and, notably, more deeply buried than Miocene rocks. Eocene strata are much closer to the basement rock, at depths where natural earthquakes occur. As of 1933, then, there appears to have been recognition among at least some leading petroleum geologists that deep drilling—drilling that extended into Eocene strata, close to underlying bedrock—was not a good idea in California.

Through the late 1920s, in any case, life remained good not only for businesses in Southern California (and elsewhere), it was also generally good for the business of earthquake science, including the small group of scientists, led by Harry Wood, who had endeavored to establish a toe-hold in Southern California. Over the next few years, as Robert Millikan continued to develop Caltech into a world-class research university, he spearheaded the hiring of one of the leading seismologists of his genera-tion, Beno Gutenberg, whose research interests focused on global earth-quakes rather than regional earthquakes or hazard. Neither Gutenberg nor his Caltech colleagues stepped into a public advocacy role, as Branner, Willis, and other Bay Area scientists had done. Charles Richter went on to become the most visible spokesman for the Seismo Lab, but along with Wood and others, he continued to focus on providing information and answering questions rather than public advocacy per se. In general, the focus of Seismo Lab research shifted increasingly away from regional earthquakes, toward global seismology. In time Harry Wood, who had worked so hard to establish earthquake monitoring and research in Southern California, eventually lamented the lab's lack of attention to local earthquake science.

Prudence came naturally to Harry Wood, a fact that contributed sig-nificantly to his success in navigating political waters to establish the Seismo Lab in the first place. He was, by all indication, a man who didn't have to be told twice to keep his mouth shut. His prudence was manifest throughout the great quake debate. For example, in January 1928, as con-cerns about Willis's prediction reached a boiling point, Wood wrote to a colleague at Stanford, mentioning a proposal that Willis had made earlier, that the Seismological Society of America hold a meeting in the Los Ange-les area that March. The proposal reflected Willis's standard modus ope-randi of seeking publicity in the hopes of furthering the cause. "Regarding the matter of holding a meeting of the Seismological Society in the south in March," Wood wrote to his colleague, "we held a conference—Millikan, Buwalda, Anderson, Thomas [a consulting engineer of considerable local importance], myself—and were agreed that the time was not propitious for a meeting here." Wood added, "Anderson was the only desenter [*sic*], and he only in a half jocular way, making the point with a grin that it would be a good thing to stir the animals up; which I think you will agree with me is not a good thing to do." In the aftermath of Willis's prediction, Wood believed more than ever that public advocacy did not serve the interests of his community any better than it served local business interests.

As talk of Southern California earthquakes fell out of the news after 1928, so too did both Hill and Willis. Although Hill yearned to return to Texas, and he did return for the visit in 1929, he was unable to find a position there. His once-warm relationship with Copper and the Building Owners and Managers Association had ended, as so many of his professional relationships had ended, with bridges burnt to a crisp. At least for a time, however, Los Angeles continued to offer some gainful employment. Hill's involvement with the St. Francis Dam tragedy had established him as a go-to local expert for geologic investigations; among other opportunities, he was tapped to serve as a consultant on proposed new dam sites in Southern California.

In early 1931, Hill did engineer a permanent retreat from California, returning at last, and for good, to Texas. Although professional opportunities in the oil industry or any other business were scarce by this time, Hill found at least modest gainful employment, reverting to his earlier roots as a newspaperman. The *Dallas Morning News*, headed by a longtime friend of Hill's, paid him twenty-five dollars a week to write a regular column on petroleum geology, focusing on the East Texas oil field. Hill moved into the Jefferson Hotel and set to work on articles describing the newly discovered massive field. Eventually stretching more than 140,000 acres and parts of five counties, the field was by 1931 the largest in the United States. In Nancy Alexander's words, Hill's articles were "no less than an extension of his lifelong attempts to establish the relationships between geology and geography and to convey to people the critical role these elements played in their culture and economy." Hill used his articles to introduce his readers to basic geology of the state as well as the East Texas field.

As time went on, the focus of Hill's articles wandered further away from geology and the East Texas oil field. He veered deeper into history, a longtime avocation, writing tales of the old town of Comanche, the Comanche of his youth. His writings also became increasingly personal and retrospective. Hill confided to his readers that there were times when he wished that he "hadn't seen that bad old frontier and had been brought up in a proper little Lord Fauntleroy manner to become a cigarette-smoking country club porch hanger." But more frequently he expressed nostalgia for his early days in Texas, before the frontier had been tamed. "Oh," he wrote in the introduction to this same article, "for a canteen of cold 'gyp'-tasting water and a saddle blanket to lie upon beneath the shade of a big mesquite tree on the edge of the prairie." In an article published six months before his death, he wrote, "I would not exchange an hour of that old-time frontier range life . . . for $10,000,000 or the finest place in Dallas."

Hill's personality shined increasingly through his articles, which offered up his views on topics as wide ranging as art, politics, religion, and Prohibition—opinions unfailingly "firm and unshakable, softened by no stroke of tact." While readers were sometimes irked by his strongly stated opinions, in increasing numbers they were drawn into and came to appreciate his world and "musings." He corresponded at some length with readers who wrote to him about the diverse topics he discussed. As Hill reminisced about the world he had left behind in 1882, California and its earthquakes faded from his thoughts. The success of his articles notwithstanding, Hill's finances appear to have remained modest at best. In an attempt to bolster his meager finances, at one point Hill ventured back into the oil business with a deal involving a new oil prospect in Coman-che. The well, eventually drilled in 1933, was dry. He did own some oil stock, presumably earned in partial payment for earlier consulting work. And from 1935 onward he received a monthly stipend of one hundred fifty dollars from the Geological Society of America. In his later years, his finances were sufficient to pay for daughter Jean's education at UCLA, a goal he had long held dear.

In the years following the refutation of his prediction, Bailey Willis also largely stepped away from the earthquake scene. For a time, Willis remained somewhat more engaged than Hill, still interested in advancing the promotion of earthquake risk reduction. Toward this end, however, he found himself with dwindling support among seismologists. Allies united on the same side of any cause might be content to swallow their reserva-tions about one another—as, in fact, Hill appears to have done when he traveled with Willis to Texas in 1898 to drum up support for future US Geological Survey mapping in Texas. During the heat of the great quake debate, Sidney Townley was another case in point. Townley, who had been active in the Seismological Society of America since its inception, includ-ing serving as longtime editor of its bulletin, had had misgivings since the mid-1920s about the extent to which Willis sought publicity. In 1926, for example, he opposed Willis's plan to garner public support among San Francisco Bay Area businessmen by publishing an article by a consulting engineer on the future growth of the San Francisco metropolitan area. In Townley's view, such editorializing had no place in a scientific journal. Townley, a professor of astronomy at Stanford who had been involved with seismology longer than Willis, had also from the beginning viewed Willis's prediction as unsound.

More generally, Townley viewed Willis's appreciation of seismology as superficial. In Townley's view, "Willis' reputation as a seismologist and a scientist is not above criticism." Indeed, Willis's letters to Harry Wood do reveal a penchant for making hay of apparent patterns in the timing of Southern California earthquakes, without careful consideration of either the full known chronology or the disparate nature of the earthquakes themselves. In 1926, even before the public refutation of Willis's prediction, Townley refused to nominate Willis to attend a scientific meeting in Japan as the official representative of the Seismological Society of America and protested privately against the insistence of others that Willis be chosen as the delegate. Other colleagues had clearly had misgivings of their own about Willis's grandstanding, as evidenced, for example, by Wood's reaction to Willis's proposal to "stir up the animals" in the Los Angeles area in 1928.

Reservations notwithstanding, until Willis's prediction had been formally refuted, Townley, Wood, and other colleagues might have had misgivings about Willis's approach, but they remained on the same team, united in their cause. Misgivings might sometimes be shared privately, but rarely pressed. As he endeavored to establish an earthquake program in Southern California, Wood corresponded extensively with Arthur Day. Most of their correspondence focused on nuts-and-bolts issues: instrumentation, seismograph sites, budgets, et cetera. When Robert Hill's statements about hazard blew up in the media in late 1927, Wood brought up the matter several times in his letters to Day, expressing concern at the extent to which hazard was being downplayed in the press. The earlier public relations dust storm that Willis had kicked up with his prediction did not, however, bear mention in Wood's letters. Wood might not have liked that publicity any better than he liked publicity downplaying seismic hazard, but the publicity that Willis's prediction generated served his own interests well enough. Once the prediction backfired, the equation changed. The public relations campaign having failed, the lead campaigner found himself with little support.

Hill himself might have viewed the publication of his book as a humiliation rather than any sort of professional success, but on the narrow issue before the court—the prediction that Los Angeles would be struck within years by a great earthquake—Hill's arguments had prevailed. Even with the benefit of hindsight, knowing about the earthquake that would strike the Los Angeles area a few years later, on the narrow question of the prediction itself, Hill's arguments had moreover rightfully prevailed. Yes, some of his scientific arguments were flawed, but to some extent his missteps were understandable for the time, and fundamentally Hill was right: Willis's prediction had been fundamentally unsound. Some if not most of their colleagues understood as much. In a letter written to Ellis Shuler in December 1927, Hill mentioned that he was sending under separate cover a copy of a speech that he had given the previous week, a speech that laid out the same refutation that he presented in his book. "This man W.," Hill wrote to Shuler, "is one of the keenest and sharpest rascals that ever disgraced science, and it was his implacable and unreasonable hatred of southerners that made it so unpleasant for me in Washington that I left it years ago. I am certainly glad to have caught him." Shuler might have brought out Hill's better angels and have himself been disinclined to dwell on disagreements, but he later responded to Hill, "I have just read your paper which you kindly left me and I certainly did enjoy it. You fixed our friend good and proper."

How Willis himself felt about the whole affair, we have few clues. He was a voluminous writer throughout his life, penning not only a great many scientific and popular articles, but also autobiographical writings and mountains of letters to his wife during his extensive travels. About the great quake debate, and Robert Hill, however, he wrote very little, including in his many letters to his wife. Written in fluid, slanting script, his letters to her were overwhelmingly bubbly, upbeat, and loving, whether he wrote about family matters or his professional activities. While he usually wrote at some length about his work, his activities in the great quake debate, and his interactions with Hill, did not bear mention through the mid- to late 1920s—at least, not in any letters that are preserved among Willis's papers. Hill does, however, make a cameo appearance in a letter he sent to Margaret on September 10, 1928. This letter describes a recent visit and wide-ranging conversation with Arthur Day. The conversation touched on the recent installation of seismometers in Southern California. Day mentioned that the Pasadena Seismology Lab had recorded more than a hundred shocks over the past year. After describing this part of the conversation, Willis wrote—his penmanship as well as his tone uncharacteristically sharp—"Oddly enough, Day did not mention R. T. H. I guess we both forgot him."

If the denouement of his prediction left Willis feeling stung, it did not extinguish his passion for the larger cause, but former allies increasingly fell away. The onetime champion of the Seismological Society of America, who had previously thrown so much of his energies into the success of the society and its cause, resigned as society president in 1927, and by 1929 he no longer served on the society's board of directors. He stepped away from the society, as it did from him, starting with the 1927 Cleveland meeting, when he instead took on increasingly active roles in the Geological Society of America.

Willis also stepped away from a public arena more generally, if not entirely. When damaging earthquakes grabbed headlines, he continued to capitalize on opportunities to advocate publicly (some might have said grandstand) for earthquake risk reduction. After the July 23, 1930, earthquake in Irpinia, Italy, brought earthquake hazard into the news with more grim stories and photographs of damage and fatalities, Willis gave an address to the National Association of Building Owners and Managers, noting that, in a severe earthquake, the "safest place in any one of our great cities will be in the steel-framed office buildings." In March 1931, about a month after a large earthquake caused extensive damage and killed more than a hundred people in the Hawkes Bay region of New Zealand, Willis wrote a lengthy article published by the *Baltimore Sun* about ongoing scientific investigations to understand earthquakes.

For all of their differences, it is ironic that in their later years, both Willis and Hill gravitated increasingly toward writing for a general audience. Both men were also interested in writing their own memoirs; in this regard, Willis, a better finisher of things in general, succeeded where Hill did not. One notes here that, throughout his life, and with at least one of his books, Willis benefited enormously from having a wife who embraced the traditional helpmeet role: Margaret collected the extensive letters that he wrote to her during an extended expedition to China, deleted the personal bits, and helped arrange to have the collection published under the title "Friendly China." In so many ways, Willis's Margaret was cut from a different cloth than Hill's Margaret.

Willis also continued to publish articles on faults, including a 1938 article on the San Andreas Fault in the *Journal of Geology* and, drawing from his 1927 investigations in the Holy Land, an article on the Dead Sea Rift in the *Bulletin of the Geological Society of America*. His days of publishing, let alone crusading, in the *Bulletin of the Seismological Society of America*, which remained the premier American journal for earthquake science, had, however, come to an end. After 1929, he published only two articles in the latter bulletin, a 1944 article on earthquakes and geological structures in the Philippines and a short correction to his earlier article

"Earthquakes in the Holy Land," which had been published in the bulletin in 1928.

Although Willis retreated from the earthquake science scene, he remained on good terms with the Carnegie Institution and other institutions. As a 1927 letter from Arthur Day to Willis reveals, Willis's activities in the mid-1920s, which had caused such consternation in some quarters, had proved critical for the US Coast and Geodetic Survey's quest to secure funding for its earthquake programs. Willis had, in effect, highlighted the critical importance of the agency's work for the earthquake problem. Throughout the great quake debate, the Carnegie Institution had remained a staunch supporter of earthquake investigations in California, pushing for an integrated program including the Coast and Geodetic Survey and Caltech. "It appears," Day wrote in 1925, "that . . . the statement you wrote for General Lord [then director of the Coast and Geodetic Survey] is the one effective asset which the Coast Survey has through which to obtain a specific appropriation for seismology."

A warm—one is tempted to say symbiotic—relationship between the Carnegie Institution and Willis continued after 1928. The institution paid for his trip to Africa in 1929, and in 1936–37 it provided support for another extended scientific expedition, this time to Japan, the Philippines, Indonesia, and India. Although Willis's letters to Margaret reveal that these long expeditions took time away from more lucrative consulting work, the grants were not insubstantial, providing support for his travel and some income. Willis pursued these opportunities actively through his seventies and even eighties. The boy who had been weaned on Cornelia Willis's tales of adventures and utterly relished his own solitary, unfettered childhood never lost his thirst for exploration or the natural world. As he had written to his mother in 1902, "Your wanderer must go to the limit of his tether." Throughout most of his adult years, his peripatetic life was supported by the devotion of a quintessential woman behind the great man. In the fall of 1940, he wrote to Margaret from a steam liner en route to a Pacific Science congress, "I look back over 42 years—thee has never been less than loyal and loving with the 'Old Fogy' to whom thee pledged thy life. And I still strive that Margy, my Pearl, shall rejoice."

The refutation of his prediction was scarcely the death knell of Bailey Willis's career, nor was it the death knell of the larger cause of earthquake risk reduction. Willis's reputation among seismologists had, however, been bruised, for having run up against powerful business interests and lost. The greater cause of earthquake risk reduction appeared to take a hit as well. In the words of author Carl-Henry Geschwind, "With the failure of Bailey Willis' public relations campaigns for greater seismic safety, earthquake researchers in California changed their tactics. Rather than trying to scare the general public, they would now cultivate alliances among professional groups that had an interest in greater building safety."

In their behind-the-scenes work, seismologists continued to find natural allies in structural engineers, who had come to appreciate the importance of earthquake-resistant design. Among the most influential engineers to join the cause was Henry Dewell, who had teamed up with Willis to publish a detailed survey of damage from the 1925 Santa Barbara earthquake and been drawn by Willis into the campaign for earthquake safety. Over the years that followed, Dewell echoed Willis's statements that large earthquakes would inevitably occur in Southern California and argued against a point of view, held by some engineers at the time, that "sound materials and honest skillful workmanship" were enough to make a building earthquake-safe. He argued that buildings in California should be designed with special provisions to ensure they could withstand lateral acceleration of 10 percent of gravity (typically shortened to g).

Here one might stop to wonder two things. Why did engineers focus on acceleration as a measure of ground motion severity? And where did the bar of 10 percent (0.1) *g* come from? Earthquakes shake the ground in complex ways: vertically and laterally, with some energy akin to the booming low tones in music as well as more jittery high tones. There are different ways to measure the severity of earthquake shaking. Traditionally, engineers have relied largely on acceleration, what people sometimes describe, not quite accurately, as "*g*-forces." Strictly speaking, *g* is not a force but, rather, the acceleration that a free-falling body will experience on earth, 32 feet (9.8 m) per second per second. The use of acceleration to measure shaking was in part practical: strong motion instruments record acceleration, and in the days before computers, it was very difficult to analyze a recording, but peak acceleration could be read from a record directly. The choice also makes sense physically: force, for example on a building, being the product of mass times acceleration.

A vertical acceleration of 1*g*—that is, 100 percent gravity—is enough to overcome gravity and throw objects into the air. An acceleration of 0.1 *g* is enough to shake small objects off of tables, but not to cause damage to a reasonably well-built structure. While strong earthquake shaking had never been recorded on a scale, by the early 1920s 0.1 *g* had been established as the strongest lateral acceleration that buildings would experience during an earthquake. This value had been estimated by engineers in Japan who investigated the catastrophic damage caused by a 1923 earthquake near Tokyo. Caltech's Romeo Martel, who had been talked back from a public statement in 1926, accepted the 0.1 *g* limit, having traveled to Japan to learn from its engineering community.

An engineer outside of California, John Freeman, became interested in earthquakes after a moderate earthquake shook New England in January 1925. By 1927 he identified a key disconnect between seismology and engineering: none of the seismometers designed to record local or distant earthquakes was capable of recording strong shaking on scale. By this time, while engineers generally agreed that earthquake shaking would not exceed 0.1 g, the estimate was based on rough calculations, not actual recorded data. Freeman urged the US Coast and Geodetic Survey to develop a so-called strong motion monitoring program in the United States. By 1932 a team of professionals from the Massachusetts Institute of Technology and the National Bureau of Standards developed compact instruments designed to record intense shaking, with a saturation level comfortably higher than 0.1 g. Early instruments were soon installed at a number of sites in California and at the newly constructed Boulder Dam, near the Nevada-California border. The focus on Southern California reveals that, at least among earthquake professionals, there remained a consensus that future damaging earthquakes were likely in the region, more so than in other parts of the United States. Then—and, to a large extent, still today—there is a strong inclination among earthquake professionals to install recording devices in the regions where they are believed to be most likely to record earthquakes.

As work went on behind the scenes, for the Pasadena Seismology Lab and business interests alike, concerns about earthquake risk soon took a backseat to more general pressures during the Great Depression. How well the strategies, in and of themselves, adopted by scientists would have worked over the long term to affect real change in a policy arena, we will never know. What nobody could have imagined in the late 1920s and early 1930s was that the earth itself was about to weigh in, in spectacular fashion, on the great quake debate, from a direction that nobody, Bailey Willis included, saw coming.

CHAPTER 11

THE CLIMAX

> For it is not the light that is needed, but fire; it
> is not the gentle shower, but thunder. We need
> the storm, the whirlwind, and the earthquake.
> —*Frederick Douglass*

On March 10, 1933, the front page of the *Los Angeles Times* carried news of the day, of which there was no shortage. Four days earlier, President Franklin D. Roosevelt had signed an emergency banking act giving him what some called dictatorial powers to take actions to deal with the continuing bank crisis that had closed the nation's banks. News from distant shores brought other news from a tumultuous time: "Nazis Unify Reich Grip," "Mussolini hails fascism's spread," "Troops of dictator in Vienna now." Not all of the news of the day was grim. At noon, "command of the United States fleet returned momentarily yesterday to the romantic days of Barry and Decatur when Admiral Richard H. Leigh, commander-in-chief, hoisted his blue four-starred flag at the towering main truck of the frigate Constitution." The newly restored ship had arrived in Long Beach as a local band played "Anchors Aweigh." The ship was expected to be a big draw; the ten-day stay in Long Beach came on the heels of a twenty-one-day stay in nearby San Pedro, during which time nearly a half million visitors "trod her historic decks." Old Ironsides was not the only ship in town; the Pacific Fleet had returned to their home base on March 9 after a six-month deployment. Across the Southland, typically pleasant spring weather conditions prevailed, with high pressure along the Southern California coast keeping nights cool and foggy.

March 10, 1933, seemed to be a day like any other in Southern California. The Great Depression had its hooks in the region, but on that spring Friday, those who had jobs went to work, and children went to school. That afternoon, sixteen-year old Tony Gugliemo, a popular and athletic boy from San Pedro, participated in a track meet at Wilson High School in Long Beach, about 25 miles (40 km) southwest of downtown Los Angeles. Following the meet, he returned to the school building to shower. At four minutes and eight seconds after five o'clock local time, an earthquake nucleated on the Newport-Inglewood Fault, about 15 miles (25 km) southeast of Long Beach. The initial rumbles felt in Long Beach would have provided the opening salvo, the first P waves racing ahead of the fault rupture itself. Although instrumental recordings of the earthquake were limited by modern standards, three different types of specialized instruments recorded the waves generated by this earthquake: sensitive seismometers operating around the globe designed to record distant large earthquakes, the half dozen seismometers that Harry Wood and his colleagues built and installed throughout Southern California to record local shocks, and three so-called strong motion instruments installed in the greater Los Angeles area just a year earlier. Combined with detailed damage observations, the data collected from these instruments have allowed modern scientists to tease out what happened after the fault began to move. The break propagated toward the northwest along the Newport-Inglewood Fault, taking aim directly at Long Beach much as the 1925 earthquake took aim at Santa Barbara and Goleta. Worse still for Long Beach, instead of tapering off gradually as the rupture neared the city, it instead appears to have picked up steam, releasing an especially energetic jolt as the rupture passed through Long Beach.

Thus did an earthquake that had nucleated in the town of Huntington Beach come to be known as the Long Beach earthquake. Within the city of Long Beach, roadways and bridges slumped, commercial masonry buildings and residential brick chimneys toppled, and, along with other local school buildings, Wilson High School sustained partial collapse. In San Pedro, Tony Gugliemo's younger sister Selma clung to her mother as the earth rocked. With no awareness of or experience with earthquakes, they believed the end of the world was upon them. Through a fearful evening and night punctuated by countless aftershocks, some substantial earthquakes in their own right, Selma and her family waited for word from Tony. Morning brought the grim news that Selma's beloved brother had been killed in the earthquake, by some accounts the only schoolchild killed in an earthquake that had been merciful in one respect, namely, striking at an hour when local school buildings were empty.

In general, "merciful" is not a word that comes to mind to describe the 1933 Long Beach earthquake. The temblor killed more than one hundred people and took an especially heavy toll on the most prevalent type of commercial construction at the time: brick and other unreinforced masonry buildings. While Long Beach itself sustained the worst damage, the temblor took an almost equally severe toll in the city of Compton, about 10 miles (16 km) north of Long Beach. Compton City Hall, a substantial block-masonry building completed just a few years earlier, sustained partial collapse. Elsewhere in Compton, some buildings pancaked and many others were damaged beyond repair. In the machine

Damage to a school building caused by the 1933 Long Beach earthquake. (USGS photograph)

shop at the nearby junior college, employees later found heavy machinery at rest on the floor, "many inches" from their original positions. "Proof that they did not slide or walk," wrote an astute observer, "was evidenced by the fact that oil pools around the feet were undisturbed." Vertical shaking in Compton had, it seems, been strong enough to negate the downward pull of gravity long enough to throw heavy machinery into the air, with horizontal shaking also moving it some distance sideways. Imagine, if you will, an earthquake strong enough not only to rock your television, but to hurl it into the air and away from the wall. Recalling Katherine Maiers's account of having been thrown off her porch and flipped over by the strongest shaking during the 1925 Santa Barbara earthquake, it is likely that, at her location, this temblor also generated accelerations upward of 100 percent *g*. The shaking from the Long Beach earthquake took a heavy toll throughout Compton—a bigger town in 1933 than Goleta had been in

1925—even to relatively resilient wood-frame houses. An unnamed motorist reported having been driving through Compton when the earthquake struck and later wrote, "I saw a two-story frame house rocking like a tree in the wind and then I saw it fall."

The day after the earthquake, the *Los Angeles Times* tallied the list of dead at 127. More than a few people died rushing out of buildings, only to be hit by debris falling from buildings that did not collapse but sustained damage to brickwork and architectural ornaments jarred loose by the shaking. Others would have rushed out if the ground hadn't been shaking so violently that they could not move. Years later, a woman who had experienced the earthquake as a child described having been at the dinner table with her family when the earthquake struck: "We tried to stand as a thunderous pounding threatened to break through the ceiling, but with the table rocking back and forth, it was impossible." When the shaking subsided, they scrambled out the back door to find the steps and porch covered with bricks; the chimney had collapsed, sending a cascade of bricks down the roof. "We suddenly realized that if we had been able to get up, we'd probably have been buried by the avalanche."

Californians today are, one hopes, familiar with the refrain "drop, cover, and hold on," the centerpiece of earthquake response messaging developed to prevent an instinctual response that, in an area where catastrophic building collapse is highly unlikely, potentially endangers rather than protects human life. There are, we now know, better and worse ways to respond when the earth starts to shake. But in any earthquake, life and death can hang on a fine balance. On Long Beach pier that evening, one young man wanted to play an arcade game "just once," stopping to do so as his possibly annoyed wife walked on. This small decision, the likes of which every person makes every day without serious consequence, made all the difference on this particular day. When the earthquake struck, the man was caught in falling debris and died the following day. His wife's decision to keep walking had taken her out of harm's way. Elsewhere, an American professor at the University of Kobe in Japan, unfortunately back in the Los Angeles area at the time, was struck on the head by a falling brick that knocked him unconscious; he survived, regaining consciousness to find himself lying next to two dead bodies, the men next to him having not been so fortunate. In Long Beach the list of dead ran the gamut, from two-year old Eddie Labago to eighty-three-year old Randoph Knaul, among a handful of earthquake victims who died of heart failure. Most of the deaths were instantaneous, including that of a twenty-eight-year old who died in a theater where he worked as an assistant manager.

But some deaths attributed to the earthquake came later. A sailor on the USS *Arizona*, survived the earthquake but died after he accidently discharged his own rifle while on patrol duty. And in Santa Monica three people, including airport manager Charles Towne, died in a plane crash the day following the earthquake; they had set off in a small plane to render aid to the quake-stricken area. On a normal day, most of these 127 sudden and unexpected deaths surely would have merited mention in the local news. The day after the Long Beach earthquake, they were statistics. The earthquake reportedly—possibly apocryphally—did save one life: an individual who had planned to take his own life with a gun, a plan that "miscarried momentarily when this evening's first shock disturbed his aim to such an extent that the bullet lodged in a lung."

The second biggest concentration of deaths outside of Long Beach was in and around Compton, where seventeen people, including Mrs. Ruby Wade and her infant son, lost their lives. Beyond the hardest-hit areas, a sprinkling of deaths attested to the smattering of severe damage that extended beyond the areas that had borne the brunt of the damage. The temblor took a human toll in towns as far away as Pacoima, where another older man died of heart failure. (Some do dispute whether such deaths are fairly attributed to earthquakes, noting that heart attack deaths are not uncommon on any given day.)

Considerable property losses notwithstanding, insurance liabilities were modest and manageable. Having been told just a few years earlier that the quake bogey had been put to rest, then hit by the economic pressures that began in 1929, many property owners had dropped earthquake insurance in the years preceding the earthquake. The damage had, moreover, not extended throughout the greater Los Angeles metropolitan area and had been limited within the city of Los Angeles itself.

Nevertheless, the 1933 Long Beach earthquake rocked Los Angeles to its core, with life and dollar losses that were impossible to ignore and, moreover, an indelibly etched understanding that a far worse bullet had been dodged. It was a long way from being a death knell for Los Angeles or even Long Beach; at an estimated magnitude 6.4, it wasn't even an especially large earthquake. Still, it was without question a wake-up call for not only science and risk reduction but also the business community and the public. Its reverberations would be felt at many levels, its impact belying its relatively modest size.

In this case, damage caused by the earthquake was in full display for the greater Los Angeles region. Overwhelmingly, while other types of construction did not escape entirely unscathed, it was clear that brick and other unreinforced masonry buildings had borne the brunt of the 1933 earthquake's destructive punch. Polytechnic High School in Long Beach provided a singular case in point: the school's separate auditorium building, a reinforced concrete structure, was hardly cracked, while other unreinforced buildings were nearly demolished. For years, scientists including Wood, Willis, and even Hill had argued that earthquake risk could be mitigated with proper construction; one could hardly have hoped for a more effective, albeit costly, demonstration than the Long Beach earthquake. So effective was the overall demonstration that, literally overnight, risk mitigation became a public cause. Newspaper articles and other publications described plans to build more-resilient cities: "The rebuilding of Long Beach and Compton probably will be done under the quakeproof building code evolved by Santa Barbara after that city's 1925 disaster, architects said."

The day after the earthquake, the *Los Angeles Times* itself, by far the most influential newspaper in the region and formerly a willing mouthpiece for the Los Angeles Building Owners and Managers Association, published a splashy front-page story. Under the headline "AN OPPORTUNITY—AND A FAILURE," the article began, "Out of the dust of disaster has sprung inspiration and opportunity. Ankle-deep in the crumbled masonry from damaged buildings in half a score of quake-shaken communities, Southern California rolled up her sleeves yesterday to rebuild and replace on modern lines the old and flimsy structures which were the chief sufferers of yesterday's temblors." Directly above this article, a large cartoon depicted a fiercely resolute woman, one arm across her bosom, the other outstretched to hold a banner: "PLANS FOR IMMEDIATE REPLACEMENT OF DAMAGED BUILDINGS WITH SAFE, MODERN STRUCTURES." An article published the next day in the *Times* further emphasized how quickly the area would rebound: "Whatever legislative relief is needed to ease conditions in the Southern California earthquake zone will be speeded through the lower house by Speaker Little, Chairman Cobb of the Ways and Means Committee and the entire Southern California delegation, with all other Assemblymen in sympathetic accord. Then it will be dumped into the Senate, where fast action is assured. The Governor will complete the cycle by signing any measure in this respect as quickly as it reaches this desk."

Front-page editorial cartoon on the morning after the 1933 Long Beach earthquake. (*Los Angeles Times*, used with permission)

Particularly indelible in the eyes of local residents after the Long Beach earthquake struck were photographs documenting the extent of damage to public school buildings, most of which had been built only recently to accommodate the rapidly growing population. In investigating deaths caused by the earthquake, the Los Angeles County coroner went above and beyond his usual role, holding a formal inquest to consider possible criminal liability of building contractors. Harry Wood was called to testify about earthquake risk. Caltech's Romeo Martel testified as well, speaking in more detail about earthquake-resistant standards. After deliberating for a week, a jury concluded that there had been no criminal negligence; the construction standards themselves had been inadequate. Although the jury found no criminal liability, the testimony and the verdict were reported widely, in local newspapers as well as construction trade journals, again underscoring the need for appropriate building standards.

Indeed, important efforts had already moved forward by the time the coroner's inquest concluded. Among the statements permanently etched in great quake debate lore, one finds the statement that, spurred by images of collapsed school buildings, the Field Act, which enacted strict standards for K–12 public school buildings in California, was passed just one month after the Long Beach earthquake. A great deal is hiding in the passive voice in which this sentence is invariably written. The Field Act did not, of course, pass itself. On March 23, less than two weeks after the earthquake, Republican California Assemblyman Don C. Field introduced Assembly Bill 2342, including the key provision: "An act relating to the safety of design and construction of public school buildings, providing for regulation, inspection and supervision of construction, reconstruction, or alteration of or addition to public school buildings, and for the inspection of existing school buildings, defining the powers and duties of the State Division of Architecture in respect thereto, providing for the collection and distribution of fees, prescribing penalties for violation thereof and declaring the urgency of the act, to take effect immediately." There is no evidence that Field had embraced the cause of risk reduction before the earthquake; his introduction of the bill had been spurred by the damage he had just witnessed, including some within his own district.

The politics that played out following the introduction of Field's bill were later described by an engineer who in 1933 served as chief assistant to the state architect. Originally intent on pressing for swift legislative action on the statewide building code, Field quickly regrouped and focused instead on a law governing school construction, having been convinced that, unlike a statewide building code, a law focusing on public school buildings was attainable and would be enforceable given the state's control over school construction. To craft legislation quickly, capitalizing on what Field and his cohorts likely realized would be a short window of opportunity, Field and his colleagues looked back to, and borrowed from, the so-called Dam Act of 1929—a state law passed after the St. Francis Dam collapse to provide strict standards for future dams in the state. Using the Dam Act as a template, Field and a small group of cohorts added regulations pertinent to school buildings, crafting the bill over a single weekend. The following day, mimeographed copies were distributed to every state senator and assemblyman. The bill passed quickly, by unanimous vote, in the assembly.

Field's bill did then encounter headwinds from prominent architects concerned that they would be forced to hire structural engineers, since few architects were qualified to handle the structural design of a building. In this instance, public sentiment aligned squarely with the cause of improved resilience. After examining damage from the earthquake, the head of the school department made a public statement that at least 6,000 children would have been killed had the earthquake struck during the school day. In one case the public took out its resentment with verbal attacks on the inspector of a damaged Long Beach school, not understanding that he had done his job properly but that the building had not been designed to resist lateral forces caused by earthquake shaking. The media rallied behind the cause as well. In advance of a senate hearing on the bill, representatives from state newspapers met with architects who had opposed the new law, telling them that "they would blast them in the headlines of the papers if they didn't go along with this bill." The act proceeded through final legislative hoops in record time, and on April 10, 1933, the governor signed the bill into law. Not only did the law provide strict standards for design and construction of public school buildings, it had teeth: the act stipulated that, whereas violation of the Dam Act was a misdemeanor offense, violation of the Field Act would be punished as a felony.

Other key risk reduction efforts moved forward apace in the aftermath of the earthquake. On March 20, the Los Angeles County Board of Supervisors adopted the building code that Santa Barbara had adopted in the aftermath of its earthquake. In April, Robert Millikan of Caltech accepted the Los Angeles Board of Education's invitation to lead a Joint Technical Committee to weigh in on future risk reduction. The report was delivered in early June, and in the words of Carl-Henry Geschwind, it was "a masterful brief on the need for greater seismic safety." The report was, moreover, quickly endorsed by key booster groups, including the Los Angeles Chamber of Commerce. If they could no longer argue that damaging earthquakes were highly unlikely in the region, they could embrace the arguments that scientists including both Hill and Willis had made all along, that earthquake risk could be mitigated.

Overall, the *Los Angeles Times* was not wrong with its front-page editorial: inspiration and opportunity did spring from the dust of disaster. As is usually the case in modern times, the Long Beach earthquake channeled resources into the stricken area. While insured losses were low, the Long Beach earthquake pried loose funds from other sources. In November 1933, for example, the state approved more than $3.5 million for the repair and rebuilding of damaged schools. More generally, the region dusted itself off and sprang back. Permits issued within just March 1933 indicate that, buoyed by reconstruction of damaged structures, the rate of investment in Los Angeles–area building more than doubled from the previous month. The rate of population growth did slow somewhat in the aftermath of the earthquake: whereas the population of the city of Los Angeles had more than doubled between 1920 and 1930, it grew less quickly, by about 20 percent, between 1930 and 1940. But at least as much as it was due to the earthquake, the slowing growth resulted from the overall economic climate, which hit the local oil industry hard. Still, even this double-whammy barely slowed the march of Los Angeles toward preeminence. In 1930, the population of Los Angeles County was just under 39 percent of the statewide population; by 1940, it was just over 40 percent. Arguably, in modern times, earthquakes tend to be agents of urban renewal. By virtue of its timing, the 1933 Long Beach earthquake was a singularly effective agent.

The phrase "teachable moment" is used to describe the window in time after a notable earthquake when people are keenly interested in earthquake science and hazard. Such windows can also be actionable moments, but action in a legislative arena never just happens. Following the 1933 Long Beach earthquake, a small group of engineers, lawmakers, and others quickly sprang into action to push forward a number of key risk-reduction actions—ironically, aided in no small measure by actions taken in the aftermath of the St. Francis Dam tragedy. The group of professionals who capitalized so swiftly and effectively on the window of opportunity created by the Long Beach earthquake did not, overwhelmingly, include the scientists who had played a leading role in the great quake debate before the earthquake struck. Science does not in and of itself reduce earthquake risk. Rather, it falls to scientists who study earthquakes and other natural hazards to understand the problem and communicate to the public and those who make policy, so that when the time comes, the moment is not only teachable, but actionable as well.

CHAPTER 12

SETTLING THE SCORE

We must not say every mistake is a foolish one.
—*Marcus Tullius Cicero*

O n the surface, as noted by Charles Richter in a rare note of editorializing in his 1958 textbook *Elementary Seismology*, the 1933 Long
Beach earthquake appeared to settle the score in the great quake debate.
"The Long Beach earthquake," he wrote, "had a number of good consequences. It put an end to efforts by incompletely informed or otherwise
misguided interests to deny or hush up the existence of serious earthquake hazard in California." Through the mid-1920s, Bailey Willis had
spoken about likely future earthquakes in Los Angeles. Among scientists,
Robert Hill had led the charge pushing back, arguing that earthquakes
were at most a manageable concern for Los Angeles and, moreover, that
there was no great menace from the Newport-Inglewood Fault. In the
eyes of the many scientists as well as the public, the Long Beach earthquake delivered a decisive verdict: Willis had been right, Hill had been
wrong. It is an interesting question to revisit: To what extent did the Long

Beach earthquake settle the score with respect to the statements both men had made before 1933?

In the eyes of the public, Willis had predicted that a major earthquake would strike the Southland within three to ten years of 1925, and an earthquake that seemed plenty major enough had struck eight years later. Initially, at least in private, Willis was quick to give himself a backhanded pat on the back, describing himself in a March 12, 1933, letter to Harry Wood as "a prophet who would rather have been mistaken." Wood quickly set the record straight, replying in a letter written the next day that the temblor was "only a moderately strong local shock which does not promise the regional relief of a major shock." Within days of the earthquake, Willis thus realized what some of his colleagues had appreciated even earlier, that his prediction had not, in fact, been borne out by the Long Beach earthquake. Willis had predicted that a major regional strain-releasing earthquake would strike Southern California—that is, an earthquake on a par with the great 1857 Fort Tejon and 1906 San Francisco earthquakes. The Long Beach earthquake was not that event. Even in 1933, scientists understood the difference between a major strain-releasing event like the 1906 earthquake, which had involved rupture of nearly 300 miles (about 475 km) of the San Andreas Fault, and moderately large temblors like those in Santa Barbara in 1925 and Long Beach in 1933.

In later media interviews, Willis did explain that the Long Beach temblor was "nevertheless not a general one and probably did not relieve the strain in other parts of southern California." In another interview, he said, "The Long Beach earthquake appears to be a shock of moderate intensity on one of the several faults of the San Pedro fault zone. This fault zone was recognized by H. O. Wood, who described it . . . in 1916." He avoided talk of prediction over the months after the Long Beach earthquake, retreating back to the familiar pitch for risk mitigation: "The disaster emphasizes the need of earthquake resistant buildings under a reasonable building code recognizing earthquake hazards." On April 18, the anniversary of the 1906 earthquake, Willis was among the speakers at a "Face the Facts" luncheon sponsored by several San Francisco civic groups, emphasizing the urgent need for more effective risk mitigation.

The Long Beach earthquake had not been the earthquake that Willis had predicted, but it had been the earthquake that Hill had long argued was so highly unlikely that residents of the Southland need not fear its occurrence. Whether or not Hill actually wrote the boldest words emblazoned on the bright red cover of his book, that the area "is . . . free from a probability of severe seismic disturbances," just as the media had not been wrong in reporting Willis's not-quite prediction as a full-fledged prediction, the misappropriation of Hill's arguments did not stray too far from words that he had said countless times in articles and speeches. Among the words in the book that Hill indisputably wrote, his dismissal of the Newport-Inglewood Fault ("it cannot be said there is any great menace") clearly did not age well.

The 1933 Long Beach earthquake belied in dramatic fashion at least some of Hill's words. But did it belie all of his arguments in his book? He had argued, for example, that the media often conveyed an inflated impression of earthquake damage. In 1933 the usual forces that shape the media's portrayal of earthquake damage were, without question, at work once again. The media and others, including scientists, rushed to photograph the most spectacular instances of damage. Of the millions of structures in the affected region, in private and public photograph collections one finds multiple shots of the same heavily damaged structures, including several school and apartment buildings and the nearly demolished three-story Continental Bakery building. The Historical Society of Long

Beach later described the Continental Bakery as the "most photo-graphed commercial . . . building in town." The bias was noted by none other than Bailey Willis, who wrote to John Merriam at Carnegie on March 31, 1933, "The Long Beach earthquake, as it will be known though the epicenter was near Newport Beach, has been greatly exaggerated, as you no doubt know."

City boosters did make some attempt to call out this bias. An aerial photograph taken sixty-six hours after the earthquake, on March 13, showed an apparently unscathed Long Beach cityscape, circulated with a notarized certification of the date the photograph had been taken. Indeed, the photograph shows the taller buildings in the city, with heights reach-ing ten stories, with no apparent signs of damage. This photograph found its way into a souvenir booklet, *Earthquake Pictures, Long Beach, California,* published in the immediate aftermath of the temblor, appar-ently with the intent of reassuring an anxious public. The cover of this small booklet shows a photograph of three armed troops, captioned "The situation in good hands." Inside the booklet, a two-page preface hit some familiar notes: "Not a single Class A building suffered any appreciable damage. . . . While the loss of a single life is to be regretted, this fatality list is comparatively small considering the total population of approxi-mately 3,000,000 who reside in the general area. . . . The manner in which various authorities assumed control and handled the various emer-gency situations immediately following the first shock offers an amazing and convincing testimonial of the calm resourcefulness of the American people." Following these arguably reasonable statements, the preface ended with a pronouncement set in boldfaced type: "The best scientific data available indicate that when a community has been severely shaken, rarely within 50 years is there another quake in the same vicinity." Ironi-cally, however, even as it attempted to assuage fears, the booklet included

Continental Bakery, "the most photographed building in town" after the 1933 Long Beach earthquake. The oil derricks in the background were in the Long Beach Oil Field, one of several fields along the Newport-Inglewood Fault that had been discovered in the 1920s. (Long Beach Fireman's Historical Museum Collection, Gerth Archives and Special Collection, California State University–Dominguez Hills, used with permission)

some of the familiar photographs of spectacularly damaged buildings, including the Continental Bakery building.

The Long Beach earthquake also seemed to mostly but not entirely bear out Hill's arguments in another respect: that structures can be resilient in strong shaking if they are properly engineered and well built. As virtually all scientists pointed out after the earthquake, including Willis and Wood among many others, damage in 1933 was largely, if not quite entirely, restricted to poor-quality unreinforced masonry buildings. In the hardest-hit regions, however, shaking was strong enough to damage not only shoddy construction but also relatively resistant wood-frame homes, reinforced-concrete buildings, and ordinary but good-quality masonry buildings like Compton City Hall. In this regard, the earthquake did belie Hill's words: "There is no evidence in all the seismic history of Southern California," he had written in his book, "of a properly constructed building having been shaken down or wrecked by an earthquake." Although he did not spell it out, these words clearly suggest that no properly constructed building would ever be shaken down or wrecked by an earthquake.

As noted earlier, in drawing some of his scientific conclusions, Hill was sometimes wrong for understandable reasons. When he argued, for example, that California was less seismically active than it had been in the distant geological past, his conclusion was based on the erroneous but common assumption of the day that faulting is primarily vertical and on the (astute) recognition that mountain-building in California is far less active in the present day than in the geological past. He was similarly wrong about the potential severity of earthquake ground motions from a local earthquake on the Newport-Inglewood Fault, for several reasons that were quite sensible at the time. First, neither he nor any of his colleagues at the time appreciated how locally severe a moderately large earthquake could be. Only decades after 1933 did scientists understand that, while a very large earthquake like the 1906 temblor impacts a much larger area than a moderately large event like Long Beach, the peak shaking from a magnitude-6.4 earthquake can, in the immediate area, rival that of a much larger earthquake.

Hill's conclusions about possible ground motions from a future earthquake on the Newport-Inglewood Fault were wrong for a second understandable reason. Recall that the best experts of the day had concluded that earthquakes would not ever generate peak ground accelerations greater than 0.1 g, a level of shaking that would not cause significant damage to even a reasonably well-built structure. Specialized strong-motion instruments, as discussed earlier, had been designed on this assumption. The fact that heavy machinery flew into the air in Compton tells us that shaking during the Long Beach earthquake had not been limited to 0.1 g but, rather, astonishingly had exceeded 1.0 g. The 1933 earthquake had generated far stronger shaking than what engineers and scientists of the day had thought possible—and not just in Compton. The nearest strong-motion instrument, installed in the city of Long Beach just a year earlier, should have provided a precise measure of ground acceleration. The instruments had been designed with what had been considered a comfortable margin for error, capable of recording shaking fully three times larger than 10 percent g; they had not been designed to stay on scale for shaking that was five or even ten times larger. The instrument in Long Beach did record the earthquake, and the vertical component was captured on scale, the first true strong-motion data ever recorded by a specialized instrument. But the shaking was so strong that the records were difficult to untangle, and the horizontal recording was off scale. Scientists now know that the most severe earthquake shaking possible tops out between 100 percent and 200 percent g, with a perhaps minuscule chance of even stronger shaking.

The shaking generated by the Long Beach earthquake belied Hill's arguments for yet another understandable reason. He had argued many times that strong shaking in any earthquake was narrowly concentrated in proximity to the causative fault. To a large extent, this argument had been well supported by observations following the 1906 San Francisco earthquake. In Hill's book, he quotes Willis himself as having said "If you are on the fault, the danger is extreme. If you are ten or twelve miles away, the danger is very materially less." While modern science generally supports this conclusion, we now know that not only can moderate earthquakes generate severe shaking, and not only can that shaking be well in excess of 0.1 g, but also that earthquakes sometimes generate very strong shaking well away from the causative fault. In any earthquake, wave propagation effects—what happens to seismic waves after they leave the fault—can serve to focus or otherwise amplify shaking locally. These effects can be complicated, dependent on how seismic waves bounce around complicated shallow structures like basins and mountains. In the 2010 magnitude-7.0 Haiti earthquake, for example, topographic amplification caused pockets of severe damage on small hills and ridges in Port-au-Prince. In 1933, three-dimensional channeling of energy combined with amplification by local sediments to hammer the city of Compton. Its relatively modest size notwithstanding, the Long Beach earthquake arguably taught scientists more about how the ground moves during an earthquake than any other earthquake in history. Hill had made some missteps with his arguments, but in a number of key respects the Long Beach earthquake was not a rebuke for him personally as much as it was a learning experience for the entire community of earthquake professionals.

In early April, just weeks after the earthquake, the Seismological Society of America convened a meeting on the earthquake at UCLA. Willis came down from Stanford, and Hill traveled to the meeting from Texas. Little has been written about this hastily arranged meeting, but it appears to have been a congenial gathering. Although Hill did don his trademark curmudgeonly mantle long enough to write to daughter Justina that "in some respects I felt like an interlocutor in the monkey show," he went on to say, "The banquet was beautiful and the meeting was harmonious throughout." Most astonishingly, he continued, "Even the great lion Bailey Willis and the lamb (poor little me) laid down happily together, so to speak." Among the character traits that Hill's papers reveal is a seeming inability of, or at least great difficulty in, accepting any portion of personal blame for his failures or difficulties, of ever admitting that he had been in the wrong. Willis, who made his own share of missteps, does not appear to have been much better in this regard. The Long Beach earthquake came along to settle scores, providing an upbraiding for both Willis and Hill, along with the scientific community in general. Both Hill and Willis had been wrong about some things, but admirably right about others.

Earthquake science can be like that. Even today, scientists still debate some important questions—for example, what the largest earthquake possible is in a given area. In Southern California, there is debate about the details of the shaking that will be produced from an earthquake on a par with the 1857 Fort Tejon temblor. Scientists know that, at a minimum, an earthquake like this one will shake a very different built environment than the one that existed back then. Some scientists have expressed concern that a repeat of the 1857 earthquake could even topple some modern high-rise buildings. Other engineers and scientists dispute this claim. And so the debates continue. In the end, in earthquake science, the earth itself eventually settles the score.

CHAPTER 13

THEATER

All the world's a stage.

—*William Shakespeare*

In the aftermath of the 1933 Long Beach earthquake, while Bailey Willis again donned the advocacy mantle and others focused on actions in a legislative arena, Robert Hill picked up his pen to write a series of articles on earthquakes that appeared in the *Dallas Morning News*. The initial article was published on March 12, less than two days after the earthquake, under the headline "Growing Pains of California Mountains Cause Earthquakes, Noted Geologist's Explanation." Hill later wrote a series of three articles on the earthquake, published in the same newspaper on April 9, 10, and 11.

Hill's March 12 article makes for interesting reading. A follower of Hill's earlier public comments on earthquake hazard in the Los Angeles region would have been surprised to read that "Southern California has all along realized that a disaster of this kind threatened it. Earthquake dangers are considered in all construction operations. It is owing to the fact that the chamber of commerce of Los Angeles has been preparing for such an event that the death list has not been greater. For years its various committees of engineers, architects, and other technical men have been preparing to avoid the conditions that caused the holocaust at San Francisco in 1906. Gas main control was provided for, architectural rules and specifications were made, and other necessary precautions taken." The article concluded on a warm note: "Neither need we fear but that Los Angeles will master its earthquake problems and recover. They believe in and use science out there to its very tip end."

Hill's later three articles in the *Dallas Morning News*, written after he had visited Los Angeles, also make for interesting reading. At times he returned to familiar refrains, downplaying the effects of the earthquake. He noted, for example, that far more lives had been lost in the 1928 St. Francis Dam collapse than in the earthquake. He mentioned the Newport-Inglewood Fault, emphasizing not the hazard associated with the fault but the "almost priceless pools of oil" found along it. In his articles he did acknowledge the seriousness of hazard, writing, for example, "If the epicenter of the disturbance had been twenty miles west of northward along the same fault line and had the shake come in school hours, I fear that Los Angeles would have real cause for worry." In contrast to the initial article in March, his April 11 article ended on a different note:

As severe as the March 10 earthquake was and as badly as the people were frightened, it is evident by the flagrant violation of safe building precautions and the number of shabby buildings constructed since warnings have been given not to do so, that Los Angeles has not yet learned the great lesson of its earthquake danger. Some day when a real first-class shake comes during business and school session hours as it may do some day then the money and pleasure-mad city will learn what real tragedy is, unless the perfectly reasonable steps of safe building are taken, which so far has not been done by the beer-mad and "gone Hollywood" people of our generation.

Throughout the series of articles, the individual sentences make sense, but they do not combine to make a coherent whole.

After the earthquake left Hill feeling (presumably) somewhat chastened, having belied at least some of his statements, he aimed for a more balanced message but fell short. Balance was never one of Robert Hill's strong suits. He countered alarmist statements effectively, more so than history gives him credit for. But where Willis's ebullient nature and, perhaps, affection for the spotlight led him to overstep in one direction, Hill's dogged and cantankerous determination led him to overstep in the other. If one can understand how Hill's inimitable personality traits contributed to this, another point remains: earthquake risk reduction lends itself to muddled messages. Every statement one starts to make can so easily be followed by "on the other hand . . . " Earthquake hazard in Southern California in particular is, as science and time have shown, high enough to be of serious concern. *On the other hand*, both science and time have also cautioned against overly breathless hype—for example, that the Big One in Southern California is "overdue." It bears repeating: the earthquake that Willis predicted—that is, a major strain-relieving event in Southern California—did not occur within three to ten or thirty or even ninety years of 1925. Scientists are surer today than in Hill's day that the lull in the San Andreas Fault will not last forever. But now as in Robert Hill's day, it remains a challenge for earthquake professionals to find the right notes and communicate them effectively enough to be heard.

But what of Hill's assertion at the end of his initial article in March 1933, the statement that appears to be so grossly at odds with his own arguments before the earthquake, not to mention the line spun by city boosters such as Charles Copper? Surely business leaders did not simultaneously decry the severity of earthquake hazard and prepare prudently for future earthquakes. Or did they? Business and community leaders certainly had pushed back hard against Willis's alarmist prediction and the burdensome increase in insurance rates it engendered. They had made an emphatic case, to Harry Wood and others, that publicity would irreparably damage both business and the public image of their city. When local leaders pulled Harry Wood aside back in 1923, they expressed support for his program, but concern specifically for publicity. "A reputation for seismicity," wrote one booster in 1927, "once established, may never be overcome; and our whole economic structure is predicated upon a continuation of growth." Throughout the great quake debate, business and city leaders had, as noted earlier, a number of legitimate concerns. They did not like what they saw as grandstanding. They did not like high insurance rates; they did not like sensationalist reporting of earthquake damage that would hurt property values and scare off much-needed capital from back East.

Throughout the great quake debate, business interests had never been entirely aligned. Insurance companies, most of which were headquartered back East, faced an urgent need to cut through the hype on both sides and understand the severity of hazard. Thus, even as local business interests in Los Angeles sought to downplay the hazard, powerful business interests outside the region had a strong motivation to support monitoring and research. Through the 1920s, much of the support for such activities came from the Carnegie Institution, at the direction of Arthur Day and, above him, John Merriam.

Even in Los Angeles, however, local business leaders had not become business leaders by virtue of being dumb people. Nor are businesspeople known in general for being blind to self-interest. When the Los Angeles Chamber of Commerce created its earthquake committee in 1920, it was interested in public relations but also in keeping tabs on the earthquake situation, including the activities of Wood and the fledgling seismology program in Pasadena. Ralph Arnold, who pops up repeatedly in our story, played a singular role in the great quake debate. By 1921 he might have been a wealthy and politically savvy businessman, but he was also John Branner's former student, with an especially keen interest in and understanding of earthquakes. In addition to business activities that made him a wealthy man, he had led scientific investigations of a couple of the early twentieth-century earthquakes in the region. In 1920 he said in a speech given to business leaders in New York that it seemed probable that the 1920 Inglewood earthquake had occurred on the "very pronounced fault that extends . . . thru [*sic*] the northwest residential district of Los Angeles." Although he went on to say that "extensive disturbances [on this fault] are not probable," in the same breath he acknowledged the "frequent evidences of seismic activity in the region."

Arnold brought his scientific expertise to the LA Chamber of Commerce earthquake committee. On September 9, 1920, he stepped in to chair a meeting when the chairman, William Mulholland, was absent. The minutes of this meeting document the discussions, including Arnold's own argument that it would be of value to collect data from the region's past earthquakes. Arnold also asked the Chamber of Commerce president, Ford Carpenter, about the cooperation of the chamber itself. Carpenter "stated that he thought it the opinion of the body that they would be glad to have the matter studied and made practical as a technical matter but that little good could be accomplished through publicity."

As the Carnegie Institution program moved forward through the 1920s, the chamber's earthquake committee also reached out to Wood on a number of occasions, inviting him to report on recent lab activities and findings. Wood echoed Hill's misgivings about Bailey Willis's prediction, but he was emphatically not inclined to downplay earthquake hazard in general. He wrote and spoke privately about his (errant but understandable) belief that, while earthquakes weren't predictable even on a decadal time horizon, the uptick in recent moderate activity since 1915 suggested that Southern California would indeed be rocked by a great earthquake sooner rather than later. During the 1920s he said as much to the Chamber of Commerce earthquake committee. In the fall of 1927, at the request of the committee, Wood prepared a statement about earthquake hazard in the area, noting that "it is strongly probable that rock strain has become acute in this region, and that a major shock may be expected in the not distant future." The committee also heard, of course, from another committee member, Robert Hill, who had spoken privately if not in his book about the need for risk mitigation.

From 1925 onward, the Chamber of Commerce earthquake committee had a twin of sorts, the Southern California Council for Earthquake Protection, that Robert Millikan established in the aftermath of the Santa Barbara earthquake. The membership of the two bodies was different, and their members moved in different orbits; they were, however, connected in the person of Ralph Arnold. Indeed, Arnold was a key link between the business and scientific communities. In March 1927, Millikan's council completed a draft of a statement on earthquake hazard and risk reduction in Southern California: "To refuse to recognize or discuss the pertinent facts, or to place upon them an unwarranted optimistic interpretation dictated by desire," the report began, " . . . can but lead to a false feeling of security." The report went on to outline a series of risk reduction methods, including building code regulation, construction standards, and city planning.

In addition to hearing the opinions of leading scientific experts, local business leaders in Southern California and elsewhere had witnessed the devastation caused by postearthquake fires in not only San Francisco in 1906, but also following the 1923 earthquake in Japan, where firestorms whipped up by an offshore typhoon swept through Yokohama and Tokyo. The humanitarian disaster in Japan far outstripped that in San Francisco, where fire had overtaken the city slowly enough for citizens to stand back and watch the conflagration. Whereas the death toll in 1906, including death from secondary causes such as suicides in the year after the earthquake, was no more than a few thousand, well over 100,000 lives were lost in Japan. An estimated 44,000 people lost their lives when a freak pillar of fire, known as a dragon twist, swept over an open area along Tokyo's Sumida River, where many had sought refuge. In the days following the 1923 Japan earthquake, Harry Wood had taken exception to a small article in the *Los Angeles Times* describing Southland temblors as "little quivers" that were nothing more than "an innocent expression of the joie de vivre." This little article was buried, however, in a sea of splashy coverage of the devastation and loss of life on the other side of the Pacific Ocean.

In the aftermath of horrific tragedy in Japan, on the heels of California's own experience in 1906, steps had indeed been taken behind the scenes, in Los Angeles as well as San Francisco, to mitigate the danger from fires following earthquakes. Among Ralph Arnold's papers, one finds a speech that is undated but refers to the earthquake a few years earlier in Japan, establishing the speech's date as 1925 or 1926. In this speech, Arnold repeated some of the familiar refrains regarding earthquake hazard in Los Angeles. But he didn't stop there. "For many years," he wrote, "the Chamber of Commerce and the newspapers of Los Angeles assumed the attitude of the ostrich and 'stuck their heads in the sand' regarding the earthquake situation. I am happy to say that this attitude, particularly on the part of the Chamber of Commerce, has changed within the last few years, and those of us who are devoting some of our time to the earthquake problem are receiving the hearty cooperation of the chamber and some of our publicity agencies." Arnold added, "The Chamber [of Commerce] has gone so far as to establish a subcommittee of seismology, the Fire and Safety Committee." He concluded, "The results already accomplished in California from the study of the situation warrant us in suggesting to the National Safety Council a policy of cooperation with other agencies throughout the United States to investigate the relationship of earthquakes to our civilization."

In the years following the earthquake in Japan, the chamber's Fire and Safety Committee also stayed abreast of work being done by scientists, reaching out to Wood among others, eager to hear his expert assessment. As Hill described in a March 12 article after the Long Beach earthquake, fire-mitigation efforts included the installation of valves to control gas mains. Other public and private agencies had taken other steps, guided by geologists from industry as well as academia. In 1925 a representative of the local power company, Southern California Edison, reached out to Arnold himself to ask for maps or literature on the locations of faults in Los Angeles. In late 1930 the chief deputy of the Los Angeles Department of Public Works wrote to Wood to express his appreciation for sharing fault maps, maps to which industry geologists had contributed substantially, as well as their academic brethren. The letter added, "It was also largely through your cooperation that we were enabled to obtain additional information from the geologists of the Union Oil Company and the Standard Oil Company, and we have been enabled to compile a map far superior to what otherwise would have been possible." With their shared keen interest in faults, petroleum geologists had long cooperated behind the scenes, to at least some extent, with their colleagues in academic circles. The fault map published by Wood and Willis in 1916 drew heavily from work by industry geologists. The cooperation continued after the Long Beach earthquake struck. In the days following the event, a Shell Oil geologist visited Harry Wood to share industry fault maps pertinent to the earthquake. Interests among the various players, from academic scientists to various facets of the business community, including the oil companies, might not have been aligned, but everyone living or working or investing in Southern California cared about earthquakes. In the words of Ralph Arnold in his mid-1920s speech, "Architects, contractors, public service corporations, and everyone in fact, are interested directly or indirectly in the earthquake problem."

Notably, the Santa Barbara earthquake had also weighed in during these years, providing its dramatic illustration of the vulnerability of unreinforced-masonry construction. This round in the great quake debate appeared to go to business interests, but behind closed doors its message was heard. And outside of California, the apparent campaign by city boosters could only ever do so much. The Santa Barbara earthquake garnered a banner headline in the *New York Times*: "SEVERE EARTHQUAKE WRECKS SANTA BARBARA; HOTELS COLLAPSE; BUSINESS AREA IS IN RUINS; 12 PERSONS KILLED; PROPERTY LOSS OF MILLIONS." The paper did publish a later article with the headline "LAYS QUAKE DAMAGE TO FAULTY BUILDINGS" and the subheader "Ralph Arnold, California Seismologist, says Santa Barbara Shock was Not Severe," but this single-column article ran a month after the earthquake, on page 25. In papers around the country, damaging early twentieth-century earthquakes in California, including but not limited to Santa Barbara, made for splashy headlines, often accompanied by photographs of especially dramatic damage.

After Willis's prediction brought the great quake debate to a full rolling boil, prudent building owners had also taken steps, in the interest of avoiding steep insurance rates, to improve construction. Clearly, steps might have gone further, but steps were indeed taken. In 1927, even as forces conspired to bring Hill's book to fruition, the California Development Association, one of the local booster organizations that had sprung up after the 1925 Santa Barbara earthquake, had appointed a committee of engineers, architects, and contractors to develop a Uniform Building Code for California, including provisions for earthquake-resistant design. The goal of the committee, which was largely led by volunteer experts, was to create construction standards sufficient to obviate the need for expensive insurance. Work on the code, which was meant to standardize all aspects of construction practice, proceeded slowly, the sense of urgency waning after Hill debunked Willis's prediction and memories of the 1925 earthquake faded. But the huge committee was also an inefficient hammer for a big—and complicated—nail. The committee nevertheless had begun work on a code, including specific provisions to make buildings earthquake safe by designing them as a unified whole, their components sufficiently tied together.

Even in the absence of a statewide building code, building owners had learned the lessons from Santa Barbara, where generally well-constructed but irregularly shaped buildings, notably the prominent San Marcos building, had along with other vulnerable buildings fared poorly. In the immediate aftermath of that event, the city of Santa Barbara formally requested the Los Angeles Chamber of Commerce's counsel to guide its rebuilding effort. In Los Angeles as well, responsible owners moved forward quietly to design more-resilient buildings. In 1931 Southern California Edison proudly unveiled its new building in downtown Los Angeles, a spectacular marriage of Art Deco design and state-of-the-art engineering. The building was one of the first all-electric buildings in the country, moreover designed to be earthquake resistant, with bracing and overall design to resist lateral forces. The building rode out the 1933 earthquake unscathed and was the venue for meetings that Bailey Willis attended, when consulting opportunities brought him back to the area in May 1933. The building survives to this day.

Other efforts had continued behind the scenes as well. In 1929 the Structural Engineers Association of Southern California was founded, in part due to interest in earthquake engineering. The very term "structural engineer" was codified in 1932 with passage of a law that the title "structural engineer" could be used by registered civil engineers who specialized in building design.

Hill's seemingly discordant words written in the immediate aftermath of the 1933 Long Beach earthquake were not, in fact, inaccurate. Through the heat of the great quake debate, business leaders and city boosters had taken steps behind the scenes to mitigate earthquake risk. If they didn't go far enough or fast enough, they did move forward. But what, then, does one make of the rounds that were fired so publicly in the heat of the debate—for example, the articles fed to the media by Charles Copper and his cohorts? Considering the work that proceeded behind the scenes on both sides, including some by the Los Angeles Chamber of Commerce itself, one can only conclude that, to a large extent, the great quake debate was about public relations. From the start, business leaders believed that nothing good would come of public talk of earthquake hazard, in particular if it was, as it tended to be, sensationalist. And so they orchestrated a campaign, show-business performances worthy of a show-business town.

On the other side of the great quake debate, Harry Wood might have been inclined to keep his head down and his mouth shut, but at least some of the key players understood the value of a good performance. In a 1923 letter to Harry Wood, Arthur Day mentioned Carnegie's decision to fund an expedition by Bailey Willis to investigate a recent large earthquake in

Chile, which had triggered a tsunami that had caused heavy damage along the coast. In his letter, Day agreed with a view Wood had expressed earlier: "That earthquake, all of the features of which were and are off the coast, does not offer outlook for tangible investigation." But Day went on to say, "Still you must bear in mind that the only time when money can be obtained for earthquake studies is when the earthquake is in people's minds." He added that Willis would be able to collect a set of information that "will make a good report and will be of very great service in our consideration of some other problems in California." Arthur Day and others at the Carnegie Institution understood that, when one is in the earthquake business, earthquakes are good for business. Day furthermore clearly realized that, for the public relations game, there was no better foot soldier than the ever-peripatetic, ever-ebullient Bailey Willis. Willis himself knew this well. In a 1940 letter to a colleague, he referred to himself as "one of the best scarers alive."

Indeed, Bailey Willis could be counted on to step onto a public stage not only mindfully but also willingly. The scientist who had been called Windy Willis by the small fry at the US Geological Survey could also be an impassioned, effective, and enthusiastic orator, one who remained deeply committed to the cause of risk mitigation. In the usual telling of the great quake debate, Bailey Willis made the decision to "scare people into action" in late 1925 with his bold prediction of future earthquakes. Looking back, one again wonders about coincidental timing and possible behind-the-scenes forces. After the 1925 Santa Barbara earthquake had grabbed but then faded from public attention, Robert Millikan had sprung into action, creating his council to advance the community's agenda. Willis's prediction had followed not too many months later. One further recalls that the resulting publicity does not appear to have been a pressing concern for Harry Wood—or, at least, not one that he felt impelled to discuss with Arthur Day, even though the two corresponded regularly and at some length. Did Bailey Willis really make the decision on his own to keep earthquakes in people's minds? Or might a wink and a nudge have propelled him onto the stage? If the Carnegie Institution, Willis's longtime patron, had objected to the role that he took on, it surely could have quickly put a stop to it. Yet Willis made his provocative public statements not once, but repeatedly.

When Willis's statements eventually caused the debate to overheat, he exited the stage willingly, refraining from even talk of earthquakes on a 'round-the-world earthquake tour underwritten by Carnegie. One suspects that, at a minimum, Willis had some inkling of the role he played in the great quake debate, a provocateur to step onto the stage to stir up the animals while his colleagues kept to their more judicious approach. If he felt stung to some extent by the refutation of his prediction and the lack of support from the seismology community, he continued to enjoy the moral and financial support of some in the community, including the Carnegie Institution, whose interests he had served well.

After Willis brought the debate center stage, responding became a redoubled concern for the business community. Business leaders fought back, even as they worked behind the scenes to take stock of earthquake hazard and take steps to reduce risk. In this campaign they found their own willing foot soldier in Robert T. Hill, a Confederate to charge into battle against the troublesome Yankee. In this role, Hill was moreover a resident of Los Angeles countering alarmist statements from a Bay Area scientist (the Mason-Dixon Line not being the only demarcation line of note in this story). Just as Willis went on to benefit from the support of organizations whose interests he had served, so too was Hill in a position to draw on support from other groups once the debate dust settled. For a time, opportunities did come his way, including consulting work on dam projects. But he could not, as Ralph Arnold counseled, let go of his battle against Charles Copper and the Building Owners and Managers Association. Charging into battle against Copper and others, with guns blazing, he burned the very bridges that, as the politically savvy Arnold surely knew, could have benefited Hill further over the years that followed.

United by a shared passion for geology, as individuals Bailey Willis and Robert Hill differed in so many ways: the adored son versus the wartime orphan, the Yankee versus the Confederate, the teenager schooled in Europe versus the young man schooled in life on the Texas frontier, the Progressive versus the Southern Democrat, the charismatic optimist versus the tactless curmudgeon, the leprechaun versus the hobbit (some might have said troll). Among the poignant differences between Hill and Willis is this: unlike Willis, and unlike city boosters, Hill never saw the whole muddled affair as theater. For Robert T. Hill, the great quake debate was mortal combat.

LEGACIES AND LESSONS

> Science, my lad, is made up of mistakes, but they
> are mistakes which it is useful to make, because
> they lead little by little to the truth.
>
> —*Jules Verne*

The great quake debate is not a new story; it has been told any number of times before. The simplest telling casts Robert T. Hill as the villain in a black hat, the geologist in the back pocket of business interests. Carl-Henry Geschwind described Hill as "the main tool [the business community] used" to fight back against Willis and his prediction. The story told by Nancy Alexander, drawing from Hill's own papers and the recollections of his daughter Justina, casts Hill as more of a victim than a villain, a man whose prudent words were co-opted, twisted, and exploited by the very business interests that had supported him. That story is also not quite right: while forces did twist Hill's words to a limited extent, it did not take a lot of twisting to make it sound like Hill viewed earthquake hazard in Los Angeles as inconsequential.

Hill's protests over his book were impassioned but not altogether coherent. It is possible if not likely that, at the age of seventy, Hill's mental faculties had started to dim. A diminution of mental faculties might explain in part the absence of a clear message in the articles Hill wrote in 1933 after the Long Beach earthquake. The muddled messages do also to some extent speak to the nature of the man. By the late 1930s, it was clear that Hill was, at a minimum, not getting any less eccentric in his later years. Nancy Alexander described one instance toward the end of the 1930s when Claude Albritton Jr., a professor at Southern Methodist University, escorted Hill back to his room at a local hotel: "It was his practice to escort Hill directly to his room . . . for fear Hill might fall if left alone. One night he and Hill were detained momentarily in the elevator as two middle-aged ladies rushed through the lobby in an attempt to catch it. Hill was aggravated by being forced to wait. The ladies finally arrived and entered with lively chatter. They were perhaps fifty years of age, of the type described by Albritton as gracious daughters of the American Confederacy, well-bred, immaculate, and scented with a mixture of very sweet, not to say overpowering, perfumes." In Albritton's words, "Hill's sight was almost gone, but his sense of smell was acute." In his later years, Hill was also "so deaf that conversation with him was difficult." With his still-keen sense of smell, Hill sniffed at the air and roared, "Claude, this hotel is full of whores! I can't see them, but I can smell them. Can you smell them, Claude?" The ladies' gay chatter ceased, leaving behind a deafening silence. Aware that a quietly whispered word would fall, literally, on deaf ears, Albritton filled the silence with soft whistling before Hill continued, "Claude, did you hear me? Can you smell the whores?" As the women froze in their places and Albritton continued to wish he was anywhere on the planet except where he stood in that moment, Hill's rant continued: "Claude, are you deaf? Answer me! Smell the whores!" After "the longest ride of [Albritton's] life," the elevator door finally opened at Hill's floor.

Hill also struggled increasingly with physical infirmities as he neared and crested his eighth decade, his earlier heart problems worsening in his later years. And as professional opportunities dried up, his financial situation waned along with health and energies. The monthly stipend of one hundred and fifty dollars that Hill received from the Geological Society of America continued to the end of his life, ostensibly to support a project focusing on a history of geologic investigation in the Southwest. The last scholarly project to occupy Hill's attentions focused on the route taken by sixteenth-century Spanish explorer Cabeza de Vaca. Reading de Vaca's accounts of geographic and geologic features, he became convinced that the explorer's route had not turned southward from Galveston into Mexico, as believed at the time but, rather, had gone through Texas, meandering through present-day Austin and San Antonio and from thence points southward.

The Geological Society of America support appears to have also been proffered in part to support Hill to write his memoirs. Hill's interests in writing these dated back to the period of despondency and introspection that followed publication of his book in 1928. With encouragement from longtime friend Ellis Shuler, Hill wrote a number of articles and eventually got as far as writing a few brief chapters. But an autobiography never came to fruition. After his death in 1941, his papers went to his daughter Justina, who made a start at crafting the material into a biography but also never completed the project. Among Hill's papers are several accounts of his early life, penned at different times for different purposes. In addition to reflecting his own perspective, the accounts do not always agree, in tone and sometimes in detail, with one another, leaving the modern

Robert Hill (left) later in life, with a colleague (undated). (DeGolyer Library, used with permission)

biographer with a mountain of autobiographical material by a not-always-consistent narrator.

In 1939 at the age of eighty-one, Hill married his third wife, Sue Johnson, but the union was short-lived: she left him after just one week, by Hill's account after she discovered that Hill's fortunes did not match his fame. The last meaningful personal relationship of his life was with a woman, Kelly Campbell, whom he employed as a secretary through his final years. Although the relationship appears to have remained platonic, Hill valued Campbell's friendship and support to the point that toward the end of his life he wrote out a will leaving his worldly belongings, including a valuable collection of books, to his "beloved and faithful secretary," cutting out the daughters who had long been the lights of his life. That will was, however, not properly executed before Hill's life ended in 1941, at the age of eighty-three. For all of his achievements and all of his acclaim, his financial circumstances at the end of the road were better than they'd been when he had stepped onto a stagecoach six decades earlier, but still modest at best.

Following his death, a longtime Texas friend wrote in a letter, "The Good Lord broke the pattern after he made Dr. Hill. As one who knew him fairly well in his later years, I can testify that . . . even his phobias were interesting." Long-standing colleagues of Hill's sometimes expressed a near-reverence for his ability to understand the geological landscape. As longtime friend and colleague—and lead suspect for authorship of the "Windy Willis" note—Charles Gould wrote, "Dr. Hill came more nearly to being a geological genius than any man I have ever known. Most of us have to work at our geology. He seemed to know it instinctively. He could travel by train or on horseback across a country, make a section of the rocks here, collect a few fossils there, and sketch an outcrop elsewhere, come back to his office and write a report better and more accurately than most of us could do after a season in the field." Another colleague described how, on a field excursion in his later years, "The old gentlemen faced west . . . and described the landscape and stratigraphy in the greatest detail, pointed to the topographic features with his cane." The colleague, Claude Albritton, of elevator ride fame, learned only later that Hill's eyesight was by that time nearly gone, having lost his vision in one eye and grown nearly blind in the other. Albritton realized that Hill could not possibly have seen the features that he described in such impressive detail. As Charles Gould had reassured Hill following the surgery on his eye, "You see more with one eye than most see with two."

Hill's innate ability to understand the geological puzzle pieces was all the more remarkable at a time when geologists did not have the benefit of aerial photographs, let alone modern technologies such as air- and space-based radar, to help sort out terrain. On both size and time scales, human beings are puny in comparison to the grand pageantry of geology. Hill's piercing insights regarding static geologic structure did not extend reliably, however, to a keen intuition of the dynamic landscape. To a large extent Hill can't be faulted, so to speak, for a limited understand of dynamic processes. In trying to understand geological forces, Hill was scarcely alone in making missteps. When Alfred Wegener first advanced the theory of continental drift in the 1920s, providing an explanation of how significant lateral stresses could be generated, one of the most vocal critics in the United States was none other than Bailey Willis. In a 1927 letter to a colleague, Willis dismissed "Wegenerism" as a "passing squall." Ironically, Willis's arguments on this matter were also wrong for understandable reasons, including that nobody at the time appreciated how dramatically the earth's climate had changed throughout geologic time. Without the plate tectonics paradigm to understand how plates are pushed around by the underlying mantle, driven by heat from the earth's core, virtually all geologists of Hill's and Willis's generation resorted to intellectual gymnastics to explain the unexplainable, including mountain-building, lateral fault motion, and volcanic islands such as Hawaii.

But if Hill's insights in understanding static landscape sometimes faltered understandably when it came to the complexities of dynamic geological processes, he failed, utterly and miserably, when it came to understanding social and political dynamics. His was a piercing, narrow genius. Where geology was concerned, his vision could not have been sharper. Yet he struggled throughout his life with interpersonal relationships. While he made and kept close friends and earned the admiration of many, difficult people and difficult situations left him flummoxed. By middle age he had gained a measure of self-awareness, but awareness of his shortcomings never translated into an ability to master them. Later in life as much as earlier, there was no quit in the man, even when sirens could not have screamed more loudly. Later in life as much as earlier, once a relationship started to go south, he dug his boot heels into the ground and drove the relationship southward to the ends of the earth. His curmudgeonly traits combined in a bad way with his freight-train tenacity, his tendency to nurse grudges, his inability to back away from unwinnable fights.

Their different roles in the great quake debate and their different temperaments notwithstanding, Hill and Willis, as scientists, in many respects could not have been more alike, united in their keen interest in and aptitude for one of the grand scientific challenges of their day. In the late nineteenth century there was an entire country to map, and both Hill and Willis were among the cadre of brilliant young geologists, with sharp minds and young legs, to whom the job fell. After Hill's death, Charles Gould described Hill as "almost the last of that galaxy of brilliant men of science who during the last two decades of the nineteenth century laid broad and deep the foundations of our science in North America," listing, among the others within this galaxy, Bailey Willis. (If Gould was indeed the author of the "Windy Willis" note, he clearly shared Hill's ability to not let his personal feelings for an individual interfere with his appreciation for his scientific contributions.)

In fact the two men had other things in common. Willis was more personable than Hill, but neither man was a politician. At least when it came to relationships with superiors, Willis was a better manager than Hill, but neither man wanted to manage. Neither had the slightest interest in building or leading an empire. From beginning to end, and at their core, both men were scientists and explorers. It fell to other sorts of men— and indeed they were virtually all men—to build empires. And by Hill's and Willis's day, there were many types of empires. We think of empire-building as an activity undertaken by businessmen and politicians, not scientists. Science is, after all, about truth, not power. But even by the time that Hill and Willis arrived on the scene, the age of the gentleman scientist was long gone; science had started to be about power as well. By the early twentieth century, the age of scientific empire-building had begun, including government agencies like the US Geological Survey, research universities like Caltech, and private institutions. We tend to know such empires as programs and to see them as different animals from business empires. But running a large program means control over large budgets and, frequently, not insubstantial personal gain in terms of prestige as well as more material rewards. Most scientists, then as now, are neither good at nor interested in empire-building. Clarence King had been tapped to lead the newly created USGS in 1879, but not being an empire-builder at heart, he found the responsibilities more onerous than the rewards were attractive and soon exited the scene. King's departure left a different sort of man, Civil War veteran John Wesley Powell, whom Willis had called a natural-born fighter, to captain the young agency. On the other side of the country, Robert Millikan set to work not too many years later to create a world-class research university very nearly out of whole cloth. Other programs were built or run by other men, including, for some time, Arthur Day and John Merriam at the Carnegie Institution.

During the 1920s, those looking to build business empires in Southern California found a willing foot soldier in Robert T. Hill. They supported him and they flattered him; if they didn't quite put words in his mouth, they orchestrated media coverage that twisted his words just enough to convey the desired message. Scientists tend to see Bailey Wills as a flawed hero but nobody's pawn. Indeed, looking back at the great quake debate, by all indications he stepped into that debate with greater awareness of his appointed role than did Hill. But from the time he entered an advocacy arena in the 1920s, his words and actions not only served, but were also driven by, outside forces. If one is looking to launch an earthquake science or risk reduction program, as Arthur Day had said in so many words to Harry Wood in 1923, in the absence of an actual earthquake, public talk of earthquake peril is the next best thing. More than once, Bailey Willis obligingly brought the risk reduction campaign to a very public stage. While more than a few of his colleagues had reservations about his approach, and his bold prediction specifically, the attention that Willis generated served well the science community and those in the community looking to build empires.

Earthquake monitoring and research programs at several institutions, including the Pasadena Seismological Laboratory and the US Coast and Geodetic Survey, fared well in the aftermath of Willis's prediction. Underwritten substantively by the Carnegie Institution, the installation of a local monitoring network began during these years, with test installations growing into a full-fledged, if sparse, seismic network by the beginning of 1932. Harry Wood continued to lead the program after the great quake debate dust settled, with increasingly close collaborations with the California Institute of Technology and a prudently symbiotic relationship with the local business community. For its part, Caltech continued to grow into a world-class research university, its Geologic Division still among the world's preeminent centers for earthquake research. The extent to which university president Robert Millikan heeded Ralph Arnold's urging to "choose men with great care" during the early years remains unclear, but from the start, the Geologic Division steered clear of a public advocacy role, and maybe other impolitic avenues.

Carnegie's Arthur Day remained close with Bailey Willis before, during, and after the great quake debate played out. If Willis's prediction got a little too much publicity, Carnegie's solution to turn the volume down—underwriting a globe-hopping scientific expedition—could scarcely have been more appealing to Willis, ever the boy scientist and explorer at heart. Carnegie continued to support him after 1933, underwriting his later scientific expeditions. Where Hill's financial circumstances were modest at the end of his life, Willis remained at Stanford, his circumstances comfortable until his death in 1949. His marriage to Margaret was a happy union from start to end, although the end was sadly abrupt. The seventeen-year age difference between the pair notwithstanding, Margaret was the one who received a dire diagnosis at the beginning of 1941, at the age of sixty-six. "The doctors," son Robin wrote to his sister in a March 19 letter, "talk in terms of 'months.'" Even the grim prognosis would prove overly optimistic: Margaret died just days later.

Willis himself lived another eight years in remarkably good health. Even through his eighties he remained slender and wiry, a still-energetic leprechaun now sporting a bushy white beard. In the summer of 1947, having celebrated his ninetieth birthday on May 31 of that year, he wrote in a personal letter, "I am busy as ever, my 90th making no difference that I can see." Among the birthday cards and notes sent by his colleagues to commemorate the occasion, one finds a letter with a warm greeting, "Dear Grand Young Man of Geology," from everyone's good friend, Ralph Arnold. The *Newsweek* article that led *somebody* to pen the "Windy Willis" note was published in June of that year. In the article, Willis did set the record straight at last regarding his supposed prediction of the 1925 Santa Barbara earthquake. The article went on to note that "His physical bearing at 90, contrasting with his classical beard, makes Willis the talk of the campus. Until recently he rode a bicycle all over the place and took strenuous workouts at the gym. Nowadays he does his setting-up exercises at home, but still walks a lot." He continued to be a familiar figure around

Bailey Willis circa 1945. (Stanford University Archives, used with permission)

the Stanford campus, even chopping wood outside his home, until nearly the very end. Willis died at the age of ninety-one following a short illness in February 1949.

In death, Bailey Willis would have one more thing in common with Robert Hill. When the memorials were written, recounting his remarkable life and many accomplishments, as they did in Hill's case eight years earlier, they overwhelmingly skipped over Willis's starring role in the great quake debate. Willis himself said and wrote little about it. If he had not fared too badly in material ways as a result of his role, for him as well as Hill, it was not a happy chapter in his life. In the eyes of the bystander outside the corridors of science, Willis's reputation had been bolstered while Hill's had taken a mortal blow. Petrie Mondell, a reporter in Los Angeles, laid out this school of thought in a February 12, 1934, letter to

Willis. The letter gushed, "When I received your letter of 7 February it was something of an event to me. In twenty years of newspaper experience, I have built up a few 'heroes' in my mind, and your name and self have been one of them. I scarcely expected to hear from the old master himself." Mondell went on to describe how, "whilst researching some recent earthquake stuff, I ran across, quite by accident, a book in the public library here by one Robert T. Hill, a geologist, who painfully staggered through a whole mass of 'evidence' discountenancing an address by you." Mondell continued that the fact that there were four copies of the book on the library shelves "struck me as sort of ludicrous, in as much as the 1933 quake came off according to your schedule, if you care to call it a 'schedule' . . . The longer the books remain on the shelves, the funnier it makes Hill seem, with all respect to him, as he probably has made a common human error."

On this occasion, Willis took the high road, more or less, with regards to his old colleague, parroting the line that Hill himself spun out: "Thanks for your comment on Hill's book," he wrote. "He and I are old friends and have not let his misfortune in having been framed by the Los Angeles realtors trouble our relations." He did however take a step down from the high road, going on to write, "The book itself is a boomerang, as you suggest." About his supposed crowning triumph of predicting the earthquake, Willis himself remained mute in his reply to Mondell and generally mute throughout his later years. Whereas Nancy Alexander described Hill as, in the end, a "man who never accepted his own worth," a paucity of self-confidence was not among Bailey Willis's shortcomings. But if Willis's failings could not have been more different from Hill's, one again returns to a commonality. In very different ways, their upbringings had instilled a deep sense of principle, principle that both men carried through their lives and their careers as scientists. This shared wellspring of principle impelled the undercurrent of mutual professional respect even in the face of abiding personal animosity. In the aftermath of the 1933 Long Beach earthquake, others might have viewed Hill as something of a laughingstock for having debunked a prediction that the earth appeared to have borne out in dramatic fashion. Especially as it became increasingly clear that the Long Beach earthquake had not been the major strain-relieving event he had predicted, at some level Willis himself must have understood that the "common human error" had been his own. He also saw firsthand, and acknowledged at least in private, how forces conspired to greatly exaggerate the effects of the earthquake, a point Hill had emphasized over the years.

Willis moreover must have known that his other popular claim to fame, having predicted the 1925 Santa Barbara earthquake, was another undeserved honor. The preliminary result indicating a rapid buildup of strain had not held up to closer scrutiny, and the earthquake was not the major strain-relieving that Willis had predicted. It also cannot have escaped his attention that former supporters in the seismological community, whose interests he had championed for so long, fell away fast after Hill dismantled his prediction.

Willis and Hill played different roles in the great quake debate, for different reasons and with different levels of self-awareness regarding their designated roles. The impacts on their subsequent professional livelihoods, along with their public reputations, were also quite different. But in terms of consequences for scientific legacy, science did and does not regard either Hill or Willis as the victor in the debate; effectively, both men lost.

If the great quake debate can be said to have any real winner, apart from the programs whose interests it served, it was later generations of Californians. Notwithstanding the heavy immediate toll of the 1933 Long Beach earthquake, it was a fortuitously timed event. The earthquake struck late enough in California's history that many people were in the area to experience and document its effects, but early enough that most of the Southland's population growth still lay ahead. Earthquake risk reduction played out differently in the Pacific Northwest, for example, where cities like Seattle and Portland grew through most of the twentieth century before scientists understood that the region faces enormous seismic hazard associated with the Cascadia subduction zone. The Long Beach earthquake was moreover arguably fortuitously timed precisely because it came so soon on the heels of the great quake debate. If Willis's prediction backfired in the end, the conflagration sparked by the prediction had brought a simmering debate out of the shadows, into a public arena, as well as front and center within a scientific arena. Local business interests had been galvanized to fight back, but also to start—quietly, behind the scenes—to come to grips with the earthquake problem. So too was the scientific community galvanized. As the debate continued, alliances were formed. Scientists increasingly found not only natural allies in engineers and architects but also allies in the insurance world and the oil industry, which had its own interests in faults and the earthquake problem. With the debate came discussion of scientific issues and a focus on the greater Los Angeles region. A spattering of moderate earthquakes between 1915 and 1932, some possibly caused by oil production, also served to fuel the debate and focus scientific interest. Modern network seismology began during these years, as Harry Wood and his colleagues installed the first-ever network of regional monitoring instruments and the Pasadena Seismological Laboratory's young assistant, Charles Richter, began analysis that would lead in a few short years to the first scale to measure the size of earthquakes.

In the course of the great quake debate, science played out the way that science is supposed to play out: unsound arguments were eventually refuted. With appropriate if sometimes messy course corrections, aided in no small part by authoritative statements from the earth itself, earthquake science as a whole moved forward. What the great quake debate illustrates is that science is a more complicated enterprise than people realize. There are absolute truths in science, but any given published study, even by an esteemed author or team, can be wrong. The scientific community as a whole can be wedded to paradigms that turn out to be wrong, and does not always embrace scientists who challenge them. In earthquake science, those who edge toward hype are more likely to be rewarded by the scientific establishment than those who suggest that hazard might be lower than estimated. And where the rubber meets the road with earthquake risk reduction, the players are far more numerous, the issues far more complex. If earthquake science is messy, earthquake risk reduction is messiness raised to a higher power.

Throughout his career Hill made cogent points about perception of earthquake risk. "I maintain," he wrote in 1920, "that fear of earthquakes is largely a state of mind, and if we allowed ourselves to be similarly affected by the far more serious and over-present menaces which surround us we would be a community of frenzied lunatics." Indeed, earthquake risk is not the worst challenge that we face as a species, or even the worst danger that menaces Southern California. Through the time of this writing, the loss of life in all US earthquakes over time is far below that caused by extreme weather events, and it is dwarfed by the number of fatalities caused by automobile accidents. Even the number of fatalities from the 1933 Long Beach earthquake was well below the toll from the 1928 St. Francis Dam collapse, and the number of fatalities from the 1906 San Francisco earthquake was less than deaths caused by the 1900 Galveston storm. Robert Hill was right about many things, including the fact that humans tend to worry about the wrong things: the potentially catastrophic earthquake versus run-of-the-mill weather events. In modern times, most now agree earthquake risk is arguably dwarfed by the existential peril posed by climate change. As Hill himself wrote more than a hundred years ago, earthquakes do not imperil the planet, nor do they imperil civilizations.

And yet, relative risk considerations aside, with their unpredictability and terrible suddenness, earthquakes surely do reach deep into our psyches, leaving us shaken in every sense of the word. In the end, we do not view it as acceptable to have one hundred or ten thousand or possibly even a million lives extinguished at once, when such losses are overwhelmingly avoidable. Larger forces and interests aside, after the earth itself lays bare the vulnerabilities of the built environment, the public demands risk reduction. Faced with stark images of earthquake destruction, people unite behind the cause: we can do better than this, we must do better than this. Largely because damaging earthquakes are infrequent in any one place, earthquake risk reduction was, is, and will always be a process. Today, as during the great quake debate, it takes crusaders and skeptics, and the business community and politicians and the media, and everyone else on this dynamic planet to move that process forward.

ACKNOWLEDGMENTS

In an age when it can be easy to believe that the sum and total of human knowledge can be accessed via the Internet, I am especially grateful and indebted to those who curate and safeguard the great wealth of material— our shared history and heritage—that cannot be found in cyberspace. This book could not have been written without the archival materials available and kind assistance of the many individuals at the libraries and archives I visited to research this book: the DeGolyer Library at Southern Methodist University, the Huntington Library in Pasadena, the Library of Congress, the Caltech Archives, and the Smithsonian Archives. I further thank the staff of the Stanford University Archives, the California State University, Dominguez Hills Gerth Archives and Special Collections, the California Historical Society, the University of Southern California, and the *Los Angeles Times* for their assistance with photographs. The book also could not have been written had a number of individuals not taken care to preserve the papers of their former colleagues and loved ones: the colleagues and descendants of Bailey Willis, Robert Hill, Ralph Arnold, and Harry Wood. Sadly, the great quake debate played out too long ago for me to meet and thank any of these latter individuals in person.

I am further grateful to my editor, Andrew Berzanskis, who reached out to me one day to ask about my possible interest in writing a book about the 1933 Long Beach earthquake. His timing was fortuitous; I had just begun to think about diving into a new project following a nearly decadelong break from book writing. If anyone wonders why a book about California earthquakes was published by the University of Washington Press, the answer is Andrew. His enthusiasm, guidance, and support brought this book to fruition. I have also enjoyed working with Neecole Bostick and am grateful for her editorial feedback. I am also grateful for the constructive feedback from two readers, who I would be happy to thank by name, but I don't know who they are. And finally, thank-you to Kris Fulsaas, for copyediting that smoothed out remaining rough edges.

Although this is not a book about science per se, along the way a couple of my colleagues were kind enough to help me understand the evolution of understanding regarding the San Andreas and other faults, as well as other bits of geology that are outside my professional wheelhouse as a seismologist. I am especially thankful to Ray Weldon for sharing his insights about the San Andreas Fault, and to my colleague Kate Scharer. I further thank Clay Hamilton, who first brought it to my attention some years ago that business interests had had a hand in the publication of Robert Hill's 1928 book, Simon Winchester for his guidance and friendship, Steve Hickman for his support for this project, and Lee Slice for thirty-eight years of support. Lastly I thank my father, Jerry Hough, who brought me to the Duke University archives to find material for a history project. It isn't every aspiring researcher who gets an introduction to archival research from a leading scholar, when she is in ninth grade.

NOTES

Several frequently cited sources are abbreviated as shown here after their first citation in full:

AYP	Bailey Willis, *A Yanqui in Patagonia* (Stanford: Stanford University Press, 1947)
CHG	Carl-Henry Geschwind, *California Earthquakes: Science, Risk, and the Politics of Risk Mitigation* (Baltimore: John Hopkins University Press, 2001)
FTG	Nancy Alexander, *Father of Texas Geology* (Dallas: Southern Methodist University Press, 1976)
IA	Bailey Willis, "Involuntary Adventures: Idlewild (1857–1867)," unpublished manuscript, Papers of BW, Huntington Library, Pasadena, CA
Papers of BW	Papers of Bailey Willis, Huntington Library, Pasadena, CA
Papers of HOW	Papers of Harry O. Wood, Caltech Archives, Pasadena, CA
Papers of RA	Papers of Ralph Arnold, Huntington Library, Pasadena, CA
Papers of of RTH	Papers of Robert T. Hill, DeGolyer Library, Dallas, TX
RTH 1928	Robert T. Hill, *Southern California Geology and Los Angeles Earthquakes* (Los Angeles: Arts Printing Co., copyright C. A. Copper, 1928)

PROLOGUE: SETTING THE STAGE

3   "Gentlemen, I want this picture": Samuel Goldwyn, *Omaha World Herald*, August 13, 1946.

4   By 1856, the population of the new state: "Population of the State," *Los Angeles Star*, June 6, 1857, 1.

5 Mr. Bell's account: H. O. Wood, "The 1857 Earthquake in California," *Bulletin of the Seismological Society of America* 45, no. 1 (1955):
45–67.

6 "the earth was shaken to its centre": *San Francisco Herald*, January 10,
1857, 1.

7 "The movement was undulating and slow": "Earthquakes in California,"
New York Herald, February 15, 1857, 1.

11 "At the same instant I saw my parents": Duncan C. Agnew and Kerry
Sieh, "A Documentary Study of the Felt Effects of the Great California
Earthquake of 1857," Appendix to *Bulletin of the Seismological Society
of America* 68, no. 6 (1978): 3.

11 "violent": B. Willis, "Earthquake Risk in California," *Bulletin of the Seismological Society of America* 13, no. 3 (1923): 92. "probably one of the
most . . .": ibid.

11 "In this and succeeding articles": ibid., 89.

12 "MANY BURIED ALIVE": *Salt Lake City Evening Telegram* 5, no. 1323
(1906): 1.

12 "CALIFORNIA A CENTER OF EARTHQUAKE ZONE": *Baltimore American*, April 19, 1906, 5.

12 headlines notwithstanding: Gladys Hansen and Emmet Condon, *Denial
of Disaster* (San Francisco: Cameron and Co., 1989), 160 pp.

13 Among the scientific community: Carl-Henry Geschwind, *California
Earthquakes: Science, Risk, and the Politics of Risk Mitigation* (Baltimore: John Hopkins University Press, 2001), 21 (hereinafter CHG).

CHAPTER 1: BAILEY WILLIS

Except where noted, direct quotes in this chapter are from the authoritative account of Bailey Willis's early years, written by Willis himself: Bailey
Willis, *A Yanqui in Patagonia* (Stanford: Stanford University Press, 1947),
152 pp. (hereinafter *AYP*).

16 "Your wanderer": Bailey Willis, letter to Cornelia Grinnell Willis, October 2, 1902, Papers of Bailey Willis, Huntington Library, Pasadena, CA
(hereinafter Papers of BW).

18 "were his only playmates": Bailey Willis, "Involuntary Adventures: Idlewild
(1857–1867)," unpublished manuscript (hereinafter IA), Papers of BW.

18 "free like Mowgli": ibid., 8. "From Mamma": ibid., 4.

20 "cast a shadow," "the shadow of lingering death," "through three agonizing years": ibid., 4. "go his individual way": ibid., 5. "called him her 'Sunshine'": ibid., 6.

20 "You have no Father more," "was troubled because," "the man lying propped up": ibid., 7.

21 "Grandfather Joseph had definite ideas": ibid., 8. "was separated from his mother," "sobbed as they drove": ibid., 11. "in the tower that," "Then I would dive": ibid., 13.

21 "It is all jolly here," Bailey Willis, letter to Cornelia Grinnell Willis, December 3, 1871, Papers of BW.

22 "it was not much of a school," "running wild like a colt": IA, 11. "said to have excellent discipline": ibid., 13. "two Englishmen," "According to their traditions," "knocked about," "the bullies got black eyes": ibid., 15.

30 "work is COLOSSAL": William Morris Davis, letter to Bailey Willis, March 9, 1894, Papers of BW.

30 "required lucid accounts": Director Charles D. Walcott, *Twenty-first Annual Report of the US Geological Survey Director to the Secretary of the Interior (1899–1900)* (Washington, DC: USGS, 1901).

CHAPTER 2: ROBERT T. HILL

Robert Hill never published his own memoirs. The authoritative biography was written by Nancy Alexander, drawing from Hill's extensive papers as well as conversations with Hill's older daughter, Justina, and close colleague Claude Albritton: *Father of Texas Geology* (Dallas: Southern Methodist University Press, 1976) (hereinafter *FTG*). Additional unpublished autobiographical materials are included among Hill's papers at the DeGolyer Library, Dallas (hereinafter Papers of RTH), including a document titled "Geological Recollections."

32 "Every happy family": Leo Tolstoy, *Anna Karenina* (1878).

33 "These battles of the Civil War": "Through the Car Window," *Dallas Morning News*, April 9, 1939, 6.

34 "ever bright and cheerful," "violent fits of temper": Jesse Hill, letter to Robert T. Hill, December 19, 1931, Papers of RTH.

35 a few faint memories: Robert T. Hill, Autobiographical Notes, Papers of RTH.

35 "muffled drums and solemn visage soldiers": "Through the Car Window," *Dallas Morning News*, April 9, 1939, 6.

35 "did not believe that ownership of slaves," "There is every reason": Jesse Hill, letter to Robert T. Hill, December 19, 1931, Papers of RTH.

36 much below average height: ibid.

36 "Our anxiety": Jesse Hill, letter to Robert T. Hill, October 19, 1931, Papers of RTH.

37 brought home from . . . Cumberland: "Through the Car Window," *Dallas Morning News*, April 9, 1939, 6.

37 "wait for the wagon": Autobiographical Notes, Papers of RTH.

37 "his own demise was inevitable": Jesse Hill, letter to Robert T. Hill, October 26, 1932, Papers of RTH.

37 "acute delusional insanity": Dr. Callendar, Tennessee Hospital for the Insane, letter to Robert T. Hill, June 5, 1885, Papers of RTH.

38 "The days at home," "long and dreary service[s]," "much bible pounding": Autobiographical Notes, Papers of RTH.

38 "The nearest substitute I ever had for a mother": ibid., annotation on genealogical records, Papers of RTH. "a bright spot in the recollection," "volley of repressive don'ts": Autobiographical Notes, Papers of RTH.

39 Jesse's support for his siblings: Jesse Hill, letter to Robert T. Hill, October 26, 1932, Papers of RTH.

39 "the prevailing northern idea that the art of speech": Autobiographical Notes, Papers of RTH.

39 "just for the wickedness," "almost fiendish delight": ibid.

40 "polluting [their] minds with disbelief": ibid.

40 pounded a rock into fragments; fossilized coral head: Ellis Shuler, unpublished remarks awarding honorary degree to Robert Hill, Papers of RTH.

40 Mrs. Sally Wood Hill obituary, *Christian Advocate*, June 4, 1893 (in Papers of RTH).

41 "never believed it was a financial burden": Jesse Hill, letter to Robert Hill, October 26, 1932, Papers of RTH.

41 "lowly profession": *FTG*, 16.

42 Little Lord Fauntleroy kind of boy: Geological Recollections, Papers of RTH.

42 "printer's devil": W. E. Wrather, "Memorial: Robert Thomas Hill," *Bulletin of the American Association of Petroleum Geologists* 25, no. 12 (1941): 2221–31; T. U. Taylor, "Robert T. Hill was a printer's devil," *Frontier Times* 14, no. 4:145–51.

42   "seldom saw butter": "Dr. R. T. Hill, Geologist, Passes Away," *Dallas Morning News*, July 29, 1941.

43   order a geology textbook: Autobiographical Notes, Papers of RTH.

43   "drummers": Wrather, "Memorial," ibid.

43   "the first inspiration," "continuous seeker of knowledge": Geological Recollections, 9, Papers of RTH.

43   "greatest loss of [his] life," "golden opportunity . . . ": Autobiographical Notes, Papers of RTH.

44   "how on earth did they get": "Pearls Cast to Swine," *Comanche Chief*, June 9, 1935 (in *FTG*).

44   "any queer or unusual kind of rock": G. A. Beeman, *Comanche Chief*, June 6, 1924 (in *FTG*).

44   "learn[ing] the language of the trail": *Dallas Morning News*, January 2, 1877.

44   "bad fortune": Geological Recollections, Papers of RTH.

44   "gambling spirit": Jesse Hill, letter to Robert Hill, May 20, 1922, Papers of RTH.

45   "stiff, dirty, hungry": Geological Recollections, Papers of RTH.

45   "where Vanderbilt University had been started," "when I revealed my desire," "came back in jeans trousers": Geological Recollections, Papers of RTH.

46   "did not know a damned thing about it": ibid. (also in *FTG*).

46   a scrap of paper: Autobiographical Notes, Papers of RTH.

46   "any person can find instruction": Ezra Cornell, speech at inauguration of Cornell University's first president, October 7, 1868.

47   "What t'ell business have we": *Comanche Chief*, March 3, 1933 (in *FTG*).

47   "be a little child again," "tailor-made clothes": Geological Recollections, Papers of RTH.

47   "A damn sight worse": Autobiographical Notes, Papers of RTH.

48   "A Negro student sat in the same class room with me": Geological Recollections, Papers of RTH.

49   "I failed to prepare myself": notation by Robert T. Hill on W. G. Hale's letter to Robert T. Hill, June 18, 1886, Papers of RTH.

CHAPTER 3: INTERSECTING ORBITS

50   "Perhaps a man's character was like a tree": attributed to Abraham Lincoln by Noah Brooks, "Lincoln's Imagination," *Scribner's Monthly*, August 1879.

51 "develop [the South's] economic resources," "Major Powell repaid them": Autobiographical Notes, Papers of RTH.

53 "Ungracious One," "a genuine specimen": ibid.

53 "He showed me a disposition": Geological Recollections, Papers of RTH (also in *FTG*).

53 According to one of Hill's colleagues: *FTG*, 47–48.

54 "give much more attention": Charles Abiathar White, letter to Robert T. Hill, August 13, 1886, Papers of RTH.

55 "Possibly you are not aware," "I am sorry," "In so far as adjudication," "You have no occasion": John Wesley Powell, letter to Robert T. Hill, March 5, 1887, Papers of RTH.

56 Robert T. Hill, "The Topography and Geology of the Cross Timbers and Surrounding Regions of North Texas," *American Journal of Science* (1880–1910) 33, no. 196 (1887): 291.

56 "It gives me great pleasure": William Gardner Hale, letter to Robert T. Hill, August 31, 1892, Papers of RTH.

56 In Arkansas, Hill reported to John Caspar Branner: Sydney Townley, "John Caspar Branner," *Bulletin of the Seismological Society of America* 12, no. 1 (1922): 1–11.

57 "loaned" to Arkansas: James D. Dana, letter to Robert T. Hill, August 31, 1892, Papers of RTH.

57 "work among the only people," "That is no place for you": Geologic Recollections, Papers of RTH.

58 "tried to use the objective method": ibid.

58 "There is but one geological laboratory": Robert Thomas Hill, "Do We Teach Geology?" *Popular Science Monthly* 40, November 1891.

59 "Hill had never taught": Thomas Ulvin Taylor, *Fifty Years on Forty Acres* (Austin, TX: Alec Book Co., 1938), 88.

59 "He was eager": "Dean Taylor Pays Tribute to Robert Hill, Geologist," *The Daily Texan*, Austin, December 16, 1934.

59 "When we lay by our icthyosauriaus": Hill, "Do We Teach Geology?" *Popular Science Monthly* 40, November 1891.

59 "Robert T. Hill was too full of energy," "He was rather hot headed": *Dallas Morning News*, January 2, 1938.

60 "He called me a liar": annotation on University of Texas file, Papers of RTH.

60 "I don't think I ever," "We may be in the woods up here": John C. Branner, letter to Robert T. Hill, November 26, 1889, Papers of RTH.

61 "were mostly of a personal character": *FTG*, 81.

61 "Hill, I'd have bought you": ibid.

61 "No man understands any other man," "I don't think," "psychological riddles," "Talk of what you have succeeded in": William Dall, letter to Robert T. Hill, April 2, 1890, Papers of RTH.

62 "confusion, from which science": Robert T. Hill, letter to E. T. Dumble, Papers of RTH. Also in J. H. Herndon, *Plea for the Life of the Geological and Mineralogical Survey of Texas and Review of the Charges Preferred against Prof. E. T. Dumble, State Geologist, for Incompetency, Plagiarism, and Maladministration in Office, and the Sham Trial Thereof* (Austin, TX: printed privately, 1891).

62 "poor Massachusetts fellow," "monkey cap," "sense enough": Papers of RTH (also in *FTG*).

62 Herndon, *Plea for the Life of the Geological and Mineralogical Survey of Texas*, ibid.; also in Papers of RTH.

63 "judge, jury, and prosecuting attorney": J. H. Herndon, letter to Robert T. Hill, May 18, 1891, Papers of RTH.

63 "Don't you think": R. S. Tarr, letter to Robert T. Hill, March 15, 1892, Papers of RTH.

63 "buried the hatchet": John C. Branner, letter to Robert T. Hill, June 2, 1892, Papers of RTH.

63 "hard to get along with": R. S. Tarr, letter to Robert T. Hill, March 15, 1892, Papers of RTH.

63 "each mention of a position": Dale A. Winkler, Phillip A. Murry, and Louis L. Jacobs, "Vertebrate Paleontology of the Trinity Group, Lower Cretaceous of Central Texas," 1–22, in *Field Guide for the 49th Annual Meeting of the Society of Vertebrate Paleontology* (Dallas: Institute for the Study of Earth and Man, 1989).

64 "received [him] with open arms," "[came] to [his] house": Autobiographical Notes, Papers of RTH.

65 "pitched into," "I was forced": Bailey Willis, letter to Margaret Willis, Papers of BW.

65 the stated purpose of the trip: Willis, *Houston Daily Post*, January 10, 1898, 6.

66   "said it would pay the state," "It will cost": "Visiting Geologists," *El Paso Herald*, January 28, 1928, 4.

66   "With a hearty greeting," "At my table": Bailey Willis, letter to Cornelia Willis, January 3, 1898, Papers of BW.

67   "but Hill, back up by a nice horse and buggy": Bailey Willis, letter to Cornelia Willis, January 9, 1898, Papers of BW.

67   "During the past week," "Hill pointed out," "Oh! The rocks and cactus!": ibid.

67   an impromptu road trip: Bailey Willis, letter to Cornelia Willis, January 21, 1898, Papers of BW.

67   "The days are full," "It is not of scenic interest": Bailey Willis, letter to Cornelia Willis, January 11, 1898, Papers of BW.

67   "Last evening Hill": Bailey Willis, letter to Cornelia Willis, January 12, 1898, Papers of BW.

68   "There is no higher authority": "The Llano Country," *Houston Daily Post*, January 18, 1898.

68   "quite a young man": "Geological Survey of Texas," *Houston Daily Post*, January 10, 1898.

68   "He intimated": "Gold in Texas," *Washington Times*, January 21, 1898, 3.

68   "This morning we are engaged": Bailey Willis, letter to Cornelia Willis, January 18, 1898, Papers of BW.

69   "Geology and Artesian Wells": *FTG*, 140.

69   "The geology of Texas": B. Willis, "Work of the US Geological Survey," *Science* magazine, August 1899.

70   "the eminent geologist": *Houston Daily Post*, January 10, 1898.

70   "I prefer the company": Robert T. Hill, letter to the Division Engineer, Chicago, Rock Island and Texas Railway (responding to unsigned letter from boyhood friend), August 28, 1902, Papers of RTH.

71   "had no knowledge of my work or problems": Robert T. Hill, letter to Alexander Agassiz, May 23, 1897, Papers of RTH.

71   "Willis, the mamby-pamby": Robert Hill, letter to John C. Branner, April 23, 1906, Papers of RTH.

71   Robert T. Hill, "Geography and Geology of the Black and Grand Prairies, Texas, with Detailed Descriptions of the Cretaceous Formations and Special Reference to Artesian Waters," part 7, in *21st Annual Report of the U.S. Geological Survey* (Washington, DC: USGS, 1901).

72   "the single largest contributor": Ellis Shuler, unpublished remarks, Papers of RTH.

74   "the land where I had to suffer": Robert T. Hill, letter to W. M. Davis, July 24, 1928, Papers of RTH.

74   "I am a very lonely," "I have an unhappy faculty": Robert T. Hill, letter to the Division Engineer, Chicago, Rock Island and Texas Railway (responding to unsigned letter from boyhood friend), August 28, 1902, Papers of RTH.

75   "noxious prohibitions," "a caress or a pat, " "became as full of buried complexes": Autobiographical Notes, Papers of RTH.

75   "That is why I have": ibid.

76   "Iowa lady," "announced that if she knew": ibid.

76   "needlessly offensive man," "hated all things Southern": Geological Recollections, p. 69, Papers of RTH.

76   "called and found Mrs. Sanford," "These are the first people": Bailey Willis, letter to Cornelia Willis, October 19, 1890, Papers of BW.

77   "valley of Virginia," "Work was the idea": Bailey Willis, letter to Cornelia Willis, October 30, 1898, Papers of BW.

78   "It is not the barking": Robert T. Hill, letter to Alexander Agassiz, November 3, 1902, Papers of RTH.

CHAPTER 4: PARTING COMPANY AND FACING DISASTER

81   "Life is either a daring adventure": Helen Keller, *Let Us Have Faith* (New York: Doubleday and Doran & Co., 1940).

81   "pungent contempt": *FTG*, 187.

81   "ransacked [his] office and papers," "what [Hill] considered a false report": Geological Recollections, Papers of RTH (also in *FTG*, 187).

81   working . . . in the Caribbean under the direction of . . . Agassiz: W. E. Wrather, "Memorial: Robert Thomas Hill," *Bulletin of the American Association of Petroleum Geologists* 25, no. 12 (1941): 2221–31.

82   "Land of Despair," "climb for water": *FTG*, 121.

82   "bold and adventurous," "ablest, probably": "Here and There," *New York Times*, October 7, 1899.

84   "vast cubes of limestone," "such perches as we could obtain," "We had abundant opportunity": Robert T. Hill, "Running the Canons of the Rio Grande," *The Century Illustrated Monthly Magazine* 61 (1901): 371–87.

84   "procuring light": ibid., 387.

84 "a picturesque account": Bailey Willis, letter to Margaret Willis, October 1, 1902, Papers of BW.

85 "There came a sort of whirlwind," "There were some eighteen vessels," "sank instantly": "Eruptions Continue," *Washington, DC, Evening Star*, May 10, 1902, 1.

85 "Thirty thousand corpses," "charred, half-dead human beings": "Thirty Thousand Bodies Lie Strewn About Through the Streets of Lava-coated St. Pierre," *St. Louis Republic*, May 13, 1902, 1.

85 "still one of the most inexplicable," "in widely distant portions": "Smouldering Furnaces," *San Diego Union*, May 18, 1902, 8.

86 "ghastly, ashen-gray," "cry of horror": Robert T. Hill, "A Study of Pelée: Impressions and Conclusions of a Trip to Martinique," *The Century Illustrated Monthly Magazine* 64 (May–October 1902), 770.

86 "daring and prolonged investigation," "Prof. Hill is the first," "The undertaking," "Prof. Hill Knows the Risks": *Fort Worth Morning Register*, May 28, 1902.

88 "one of whom kindly placed": Hill, "A Study of Pelée," ibid., 777. "dim flare of light": ibid., 778.

88 "Following the salvos": "Volcano's Moods," *Idaho Statesman*, May 29, 1902, 1.

88 "so fast that it was easy for me to see": Hill, "A Study of Pelée," ibid., 778.

88 "bungled": ibid., 779.

88 "Mysteriously and silently": ibid., 779.

88 "noted explorer": "Fear Noted Explorer Has Been Whelmed," *San Jose Evening News*, May 28, 1902, 1.

89 "The sides of which": Hill, "A Study of Pelée," ibid.

89 "Friends of the noted explorer": *San Jose Evening News*, May 28, 1902, 1.

89 "completely worn out," "My attempt to examine," "near the ruins," "While these eruptions continue," "but [I] do not hesitate": "Volcano's Moods," *Idaho Statesman*, May 29, 1902, 1.

89 "Our greatest problem": "Brought Back from Mt. Pelee," *New York Times*, June 15, 1902, 25.

89 "*nuée ardente*": Alfred Lacroix, *La Montagne Pelée et ses eruptions* (Paris: Masson et Cie, 1904), 662 pp.

90 "HE DARES DEATH": "He Dares Death to Aid Science," *Atlanta Constitution*, May 28, 1902, 3.

90 "a man of wonderful vigor": Henry F. Beaumont, "A Tennesseean Who Won Fame," *Atlanta Constitution*, August 17, 1902, D5.

90 "When I endeavor": Hill, "A Study of Pelée," ibid.

93 "miserable and lonely": Robert T. Hill, diary, August 30–September 18, 1898, Papers of RTH.

93 "last degradation": Hill, diary, June 6–7, 1903, Papers of RTH.

94 "I had five years," "as I have done": Robert T. Hill, letter to Charles Walcott, July 1, 1898, Papers of RTH.

94 "the well-known fact": *FTG*, 194.

94 "While I admit": Robert T. Hill, letter to Charles Walcott, July 1, 1898, Papers of RTH.

94 "the work has been unsatisfactory to you": Charles Walcott, letter to Robert T. Hill, June 18, 1904, Papers of RTH.

96 "one of the most eminent," "offices in New York," "As is well known," "He is said to have made more": *Corpus Christi Caller Times*, February 12, 1904, 6.

96 "The inexplicable Jinx": Robert T. Hill, diary, November 15–16, 1908, Papers of RTH.

96 "I do not propose": Alexander Agassiz, letter to Robert T. Hill, May 10, 1907, Papers of RTH.

97 "great First Cause": Robert T. Hill, letter to Ellis Shuler, December 18, 1927, Papers of RTH.

97 taint of illegitimacy: *FTG*, 218.

98 "a bit of autobiography": *AYP*, iii.

99 "The bottom had fallen out": *AYP*, 148.

99 "Miss Baker": Bailey Willis, letters to Margaret Baker, 1894, Papers of BW.

99 "Marjorie": Bailey Willis, letter to Margaret Baker, summer 1897, Papers of BW.

99 "Dear little wife (to be)," "Lovingly, Your Old Fogy": Bailey Willis, letter to Margaret Baker, late summer 1897, Papers of BW.

CHAPTER 5: GOLDEN STATE

101 "The chief line of study": Robert T. Hill, "The Rifts of Southern California," *Bulletin of the Seismological Society of America* 10, no. 3 (1920): 149.

101 "the 29th and final draft": Robert T. Hill, diary, 1912, Papers of RTH.

104 "hated all things southern": *FTG*, 187.

104 "slow at ideas": Bailey Willis, letter to Cornelia Willis, October 30, 1898, Papers of BW.

104 Working alongside Branner (Willis's involvement in SSA): CHG, 47.

106 investigations of known oil fields: Charles Walcott, letter to Bailey Willis, July 14, 1917, Papers of BW.

106 she took drugs: *FTG*, 220.

107 concerned that Jean might inherit her mother's mental illness: *FTG*, 236.

108 "irascibility," "Wilkins curse," "Psychologically," "should not be treated": Robert Hill, letter to Jesse Hill, May 20, 1922, Papers of RTH.

108 "I wonder if the children": hand-typed quotation with note attributing it to Adela Rogers St. John (original source unclear), Papers of RTH.

109 "sane periods," "There was never a case," "Do you realize": Jesse Hill, letter to Robert T. Hill, May 20, 1922, Papers of RTH.

109 "Not a false pride," "It is because of this," "made Tom a dreamer," "Our father," "not believe that the gambling spirit": ibid.

111 "Earthquakes, as terrible as they may seem," "These processes are not destructive": Robert T. Hill, "The Mystery of Earthquakes," *Arkansas Gazette*, January 10, 1909, 1.

111 "Earthquakes are the least harmful": Robert T. Hill, "The Rifts of Southern California," *Bulletin of the Seismological Society of America* 10, no. 3 (1920): 146.

112 "In view of the vastly more": ibid., 146.

112 "secondary" but "physiographically quite conspicuous": ibid., 148.

113 "In Southern California this rift is," "In Southern California the San Andreas," "evidence of quite recent movement," "In general": ibid., 147. "In conclusion," "All I can say": ibid., 149.

114 "most interesting and wonderful": ibid., 149.

114 "In conclusion, I cannot help repeating": ibid., 149.

115 "Fear is the cruelest of all devil": Robert T. Hill, speech to Los Angeles Rotary Club, ca. 1920 or later, Papers of RTH.

CHAPTER 6: FRAMING THE DEBATE

117 "The happy impute all their success": Jonathan Swift, *The Works of Jonathan Swift* (London, Henry G. Bohn, 1843).

121 "Seismology owes a largely unacknowledged debt": Charles Richter, interview by Henry Spell, 1978, transcript, 162 pp, Caltech Oral Histories, Caltech Archives, Pasadena, CA.

121 Harry O. Wood, "California Earthquakes: A Synthetic Study of Recorded Shocks," *Bulletin of the Seismological Society of America* 6, no. 2–3 (1916): 55–180.

121 Harry O. Wood, "The Earthquake Problem in the Western United States," *Bulletin of the Seismological Society of America* 6, no. 4 (1916):196–217.

121 "There is overwhelming geologic evidence," "Therefore it must be clearly recognized," "In justice to the future development": ibid., 198.

124 "TOTAL TEMBLOR LOSS": *Los Angeles Times*, April 12, 1918, 1.

124 "TRUTH ABOUT EARTHQUAKE": ibid.

124 "The great Los Angeles aqueduct": ibid.

124 "EAST ALARMED; IS REASSURED": ibid.

125 "Many telegrams reaching": ibid.

125 "One paper indulged in a tremendous display": *Golden Age*, October 15, 1865.

126 Nepal's cultural heritage had been destroyed: for example, see "Nepal's Archaeological Heritage Destroyed in Earthquake," *Denver Post*, May 2, 2015.

126 "EARTHQUAKE DOES HEAVY DAMAGE": *New York Times*, April 22, 1918.

128 "Additional evidence": Stephen Taber, "The Inglewood Earthquake in Southern California, 21 June 1920," *Bulletin of the Seismological Society of America* 10, no. 3 (1920): 141. "The Inglewood-Newport-San Onofre fault": ibid., 143.

129 conclusions in another article: Stephen Taber, "The Los Angeles Earthquakes of July, 1920," *Bulletin of the Seismological Society of America* 11, no. 1 (1920): 63–79. "there is every reason": ibid., 77. "of utmost importance": ibid., 78.

130 "afraid to say [the word] earthquake out loud": CHG, 51.

130 "queer things": John C. Branner, letter to Homer Laughlin Jr., August 6, 1920, Papers of John C. Branner, Huntington Library, Pasadena, CA.

130 "Earthquakes of Southern California variety are much akin": "The Earthquake," *Los Angeles Times*, April 23, 1918, I14.

130 created an earthquake committee: Ralph Arnold, letter to Robert T. Hill, October 13, 1920, Papers of RTH.

131 "Hard temblor at Inglewood," "No Danger Here of Earthquake," "We had an adjustment in 1918," "I have observed no great settlings," "There will be no great damage": *Hemet News*, June 25, 1920, 2.

131 drill for oil in the Painted Hills: *Riverside Daily Press*, August 10, 1921, 4.

132 "Los Angeles is situated": Andrew C. Lawson, letter to E. W. Bannister, February 23, 1927, copy in Papers of Harry O. Wood, Caltech Archives, Pasadena, CA (hereinafter Papers of HOW).

132 "boosterism": Judith Goodwin, *Millikan's School: A History of the California Institute of Technology* (New York: W. W. Norton, 1991), 132.

132 "signs of reawakening seismic activity": Harry Wood, letter to John Merriam, September 14, 1921, Papers of HOW.

132 the Carnegie Institution agreed to provide support for the program: CHG, 57.

133 "as a fact-finding committee": *Riverside Enterprise*, August 14, 1921.

134 "Dr Arthur L. Day . . . feels," "For our part . . . we have never regarded": "Hollywood and Earthquakes," *Los Angeles Times*, September 7, 1923, I14.

134 "If that is a fair sample": Arthur Day, letter to Harry Wood, September 19, 1923, Papers of HOW.

135 "carried [him] off," "some interesting and influential people," "Two of these men," "They advised me," "This is my own judgment emphatically": Harry Wood, letter to Arthur Day, October 4, 1923, Papers of HOW.

135 "A wealthy and philanthropically inclined": ibid.

135 the Pasadena Seismological Laboratory was established: "To Locate Quakes," *Los Angeles Times*, June 3, 1925, A8.

136 Willis arranged for his son Robin: CHG, 64.

CHAPTER 7: AT THE EPICENTER

138 "Suggest something": Virginia Hugill, "Tourist Can't Elude California's Charm," letter to *Los Angeles Times*, June 14, 1925, C2.

139 "His retirement," "In California": Eliot Blackwelder, memorial to Bailey Willis, Biographical Memoir, 20 pp., National Academy of Sciences, Washington, DC, ca. 1949.

139 Neal co-authored an article: R. N. Ferguson and C. G. Willis, "Dynamics of Oil-field Structure in Southern California," *Bulletin of the American Association of Petroleum Geologists* 8, no. 5 (1924): 576–83.

139 "the other geologists": Bailey Willis, letter to Margaret Willis, June 28, 1925, Papers of BW.

140 a deep test well: "Deep Test Well at Summerland Holds Interest," *Los Angeles Times*, May 25, 1925, 14.

140 "The writer was at the hotel Miramar": Bailey Willis, "The Santa Barbara Earthquake of June 29, 1925," *Bulletin of the Seismological Society of America* 15, no. 4 (1925): 255–78.

140 "Recognizing the meaning of the shock," "Had the motion continued": ibid., 264.

141   a civil engineer by the name of Fitzgerald: Bailey Willis, letter to Harry Wood, July 3, 1925, Papers of HOW; Bailey Willis, letter to Mrs. W. B. Meloney, December 16, 1930, Papers of BW.

141   "dressed without haste," "the nails pulled in and out": Bailey Willis, "The Santa Barbara Earthquake of June 29, 1925," *Bulletin of the Seismological Society of America* 15, no. 4 (1925): 264.

141   "Going downstairs," Bailey Willis, letter to Harry Wood, July 3, 1925, Papers of BW.

141   "All the chimneys were thrown down": Bailey Willis, "The Santa Barbara Earthquake of June 29, 1925," *Bulletin of the Seismological Society of America* 15, no. 4 (1925): 264.

141   "As I stepped from the front porch," "At the instant," "more firmly bonded structures," "as though some one had run into": ibid.

142   "There has been little in the papers," "[they] did not have a business section to lose," "To give you an idea": Katharine Maiers, letter to Bailey Willis, July 10, 1925, Papers of HOW.

144   "Another thing," "There is another suggestion": ibid.

146   "New Quake Coming in about 15 Years, Rotary Club Told," *Palo Alto Times*, December 18, 1923, 1.

146   Privately he did explain: Bailey Willis, letter to Harry Wood, July 3, 1925, Papers of BW.

147   As he noted: Bailey Willis, letter to Margaret Willis, June 28, 1925, Papers of BW.

147   address a gathering: Bailey Willis speech, San Francisco Chamber of Commerce, July 2, 1925, Papers of BW.

147   A month later: Bailey Willis speech, "Face the Facts" luncheon, San Francisco, August 1925, Papers of BW.

147   "unnatural faults": Bailey Willis, "Faults (Natural and Unnatural)," *Bulletin of the Allied Architects Association of Los Angeles* 1, no. 10 (1925): 1.

147   Wood's article: Harry O. Wood, "The Practical Lessons of the Santa Barbara Earthquake," *Bulletin of the Allied Architects Association of Los Angeles* 1, no. 10 (1925): 2–3.

148   This view of an earthquake cycle: J. L. Hardebeck, K. R. Felzer, and A. J. Michael, "Improved Tests Reveal that the Accelerating Moment Release Hypothesis Is Statistically Insignificant," *Journal of Geophysical Research: Solid Earth* 113, no. B8 (2008). doi:10.1029/2007JB005410.

148   an introductory paragraph: CHG, 77.

148   "now [be] one of the safest places": Bailey Willis, press release printed as an article, "Poor Building Is Blamed," *Portland Oregonian*, July 2, 1925.

149   "too rambling and incoherent": CHG, 90.

150   "I know that my heart is in terrible shape": Robert T. Hill, letter to Ellis Shuler, 1923, Papers of RTH.

150   "rivers like men become fixed": Robert T. Hill, "Summary of Physiographic Investigations Made in Connection with the Oklahoma-Texas Boundary Suit," *University of Texas Bulletin*, July 15, 1923, 157–71.

151   "We had a hideous earthquake out here": Robert T. Hill, letter to Ellis Shuler, July 5, 1925, Papers of RTH.

151   "a great book," "the story of the Los Angeles Region": Robert T. Hill, letter to Ellis Shuler, November 6, 1925, Papers of RTH.

151   "monomaniacally employed": Robert T. Hill, diary, mid-December 1925, Papers of RTH. "with a clear conscience": ibid., January 1, 1926, Papers of RTH.

152   "They hustled me out," "It uses up all of my pep": Robert T. Hill, letter to Ellis Shuler, October 24, 1926, Papers of RTH.

CHAPTER 8: THE PREDICTION

153   "Occasionally a professional man": Charles Richter, unpublished notes, 1976, Papers of Charles Richter, Caltech Archives, Pasadena, CA.

154   "local isolated slip," "reconstruction has already started": CHG, 74.

154   "Splendid, Santa Barbara," "more than thirty American cities," "So far as earthquakes are concerned": *Los Angeles Times*, July 2, 1925, A4.

154   "largely to acquaint inhabitants": "Tells East Real Cause of Quakes," *Los Angeles Times*, July 26, 1926, B1.

155   "jerry-building was responsible": Ralph Arnold, speech at Columbia University, 1925, Papers of Ralph Arnold, Huntington Library, Pasadena, CA (hereinafter Papers of RA).

159   12 percent increase in total property valuation: "Santa Barbara Realizing Dream of City Builders," *Los Angeles Times*, February 6, 1927, B1.

159   "The East or the Mississippi Valley": *New York Times*, July 24, 1925, 25.

159   "I think it is well-written," "blamed all of the deaths": Ralph Arnold, letter to A. G. Arnoll, July 28, 1925, Papers of RA.

160   "[Willis] decided to embark": CHG, 84.

161   "At what time future shocks": Harry O. Wood, "The Earthquake Problem in the Western United States," *Bulletin of the Seismological Society of America* 6, no. 4 (1916): 200.

161   "No one knows": *Daily Palo Alto*, November 1925.

162   "Los Angeles, or its immediate vicinity," "stated three years ago": "Prof. Willis Predicts Los Angeles Tremors," *New York Times*, November 4, 1925, 9.

163   "I send you herewith": A. G. Arnoll, letter to Ralph Arnold, November 30, 1925, Papers of RA.

164   "I regard as probable": Bailey Willis, speech to National Board of Fire Underwriters, New York City, May 1926, in Papers of BW.

164   survey of some 2,700 commercial buildings: CHG, 87.

165   "I wonder if you have any idea": ibid., 89.

167   "a strong geologic department," "I have in mind": Ralph Arnold, letter to acting president of Caltech, November 1920, Papers of RA.

168   "You may take some satisfaction": John Buwalda, letter to Robert T. Hill, Papers of RTH.

168   distinguished fact-finding committee: *Riverside Enterprise*, August 14, 1921.

169   "One would judge," "We suggest": *Santa Barbara Morning News*, February 11, 1926.

170   "I like to remember," "Your father was waiting," "What a man!": Payson Treat, letter to Margaret Willis, March 1949, Papers of BW.

170   "Look who's here!," "An honestly descriptive heading": unsigned undated note, Papers of RTH.

171   Hill's papers remained: *FTG*, xi.

171   Thomas Wayland Vaughan: Thomas G. Thompson, *Thomas Wayland Vaughan, 1870–1952, A Biographical Memoir* (Washington, DC: National Academy of Sciences, 1958), 399–437.

171   "Hill was generous with his ideas": T. Wayland Vaughan address at Hill Anniversary Dinner of the Branner Club, January 14, 1927, in Papers of RTH.

172   "a young geologist": Bailey Willis, letter to Margaret Willis, October 1902, Papers of BW.

172   Gould's penmanship: Charles Gould, notes to Robert Hill, Papers of RTH.

174 Carnegie approved the grant: Bailey Willis, letter to Arthur Day, December 1926, Papers of BW.

174 "even on earthquakes": Bailey Willis, letter to Margaret Willis, January 11, 1927, Papers of BW.

175 "He is one of the most able": P. S. Smith, acting director of the US Geological Survey, letter for anniversary commemoration of one of Hill's papers, December 1926, reprinted on back dust cover of Robert T. Hill, *Southern California Geology and Los Angeles Earthquakes* (Los Angeles: Arts Printing Co., copyright C. A. Copper, 1928) (hereinafter RTH 1928).

175 "compliments heaped on compliments": *FTG*, 243.

175 "He has been hailed," "contributed largely to the selection": "Geologists in Tribute to Dr. Hill," *Los Angeles Times*, January 14, 1927, A8.

175–76 "an eminent geologist," "There is not a thread of evidence," "Accustomed as I am," "Most of the faults or rifts": "Geologist Discounts Earthquake Fear," *Los Angeles Times*, July 14, 1927, A11.

176 "Do not get the idea": Robert T. Hill, speech given to Building Owners and Managers Association, copy in Papers of HOW.

176–77 "of international repute," "prophecies of Bailey Willis," "there is no real need": "City Found Safe from Temblors," *Los Angeles Times*, December 2, 1927, A11.

177 "There is now in course of preparation": "Geologists in Tribute to Dr. Hill," *Los Angeles Times*, January 14, 1927, A8.

CHAPTER 9: THE BOOK

178 "But, my child": Ecclesiastes 12:12 (New Living Translation).

180 his presentation focused, "It is to be hoped": Robert T. Hill, "Earthquake Conditions in Southern California," presentation at the Geological Society of America meeting, December 1927, Cleveland, Ohio, published in the *Bulletin of the Geological Society of America* 39 (1928): 188–89.

180 "Brief remarks were made": ibid., 189.

180 "was still unconvinced": Robert Hill, unpublished document regarding alleged alterations in his book, p. 20, Papers of RTH.

180 "friendly calls," "no judgements": ibid.

181 "historically Southern California has been free": "Temblor Talk Tumbled Over," *Los Angeles Times*, January 19, 1928, 1.

181   "with an ok as to": Robert Hill, unpublished document regarding alleged alterations in his book, p. 22, Papers of RTH.

181   "You can see now," "don't send me anything": Charles A. Copper, letter to Lloyd Wright, August 2, 1928 (RTH letter itself not found), Papers of RTH.

181   "hideous and untruthful publicity": *FTG*, 247.

181   "endorsement," "completely [abandoned]," "elaborate arguments," "only the most perfunctory response": C. A. Copper, "Scientific World Backs Dr. R. T. Hill's Findings on L.A. Quake Situation," *Los Angeles Commercial and Financial Digest*, February 1, 1928, 8.

182   "If you are in a position": A. L. Lathrop, letter to US Geological Survey Director George Otis Smith, March 23, 1928, printed in two-page folio distributed with RTH 1928.

182   "It is true that," "In so far then": USGS Director George Otis Smith, letter to A. L. Lathrop, March 31, 1928, printed in two-page folio distributed with RTH 1928.

184   "Florida has been mighty nice," "We have great sectional rivalries": Will Rogers, letter published in numerous newspapers, including "Red Cross on Job as Usual, Says Will Rogers," letter to editor, *Detroit Times*, March 15, 1928, 33.

185   J. D. Rogers, "A Man, a Dam, and a Disaster," *Southern California Quarterly* 77, no. 1–2 (1995): 1–109.

185   "Since no haste," "These had been printed": Robert Hill, unpublished document regarding alleged alterations in his book, Papers of RTH.

186   "This book completely refutes," "Here are first given": dust jacket, RTH 1928. "Dr. Hill has had one of the most distinguished careers": ibid., back jacket.

187   "against quake doctor publicity": *FTG*, 248.

188   "Since I believe," "there is no reason for expecting": RTH 1928, 2.

188   "cornerstone," "If this allegation . . . should prove": ibid., 8. "That southern California has long been," "These phenomena have been considered": ibid., 9.

189   "There is much evidence": ibid., 41.

189   "There is no proof," "mild seismicity": ibid., 27.

191   "Such meager data": ibid., 49. "fundamental idea": ibid., 52. "This book completely refutes": ibid., front dust jacket. "Thus the foundations upon which": ibid., 62.

191 "Do not get the idea that I underestimate": Robert T. Hill, speech given to Building Owners and Managers Association, copy in Papers of HOW.

192 "[had] butted into a situation": Robert T. Hill, letter to Ellis Shuler, undated, Papers of RTH.

192 "Why God Almighty put," "I am very fond of": ibid.

193 "The predictions made several years ago," "has the style of a legal brief," "glaringly false," "In justice to the author," "One who knows Hill's other publications": William Morris Davis, "The Earthquake Problem in Southern California," galley proof, 1928, in Papers of RTH, box 34.

194 "This little book," "which should enable even the layman," "If we use even ordinary efficiency": Ralph Arnold, "Expert Studies Quake Problem," *Los Angeles Times*, April 29, 1928, B7.

194 "Expert Shoos Quake 'Bogey'": *Los Angeles Times*, April 16, 1928, 1.

194 "This book completely refutes the predictions . . . that Los Angeles is about to be destroyed by earthquakes": RTH 1928.

194 "This book completely refutes the predictions . . . that Los Angeles is endangered by earthquakes": *Los Angeles Times*, April 16, 1928, 1.

195 "this area is not only free from a probability of seismic disturbances": ibid.

195 "severe seismic disturbances": RTH 1928, front dust jacket.

195 "certainly the longer distance": RTH 1928.

196 "numerous local organizations": "Expert Shoos Quake 'Bogey,'" *Los Angeles Times*, April 16, 1928, 1.

196 "the high rates are traceable," "divergence of opinion," "no anxiety among financiers": "Quake Burden Lifts in State," *Los Angeles Times*, September 24, 1928, A1.

196 "It seems hard": ibid.

197 "Coming around to a realization": Robert T. Hill, diary, March 7, 1928, Papers of RTH.

197 "damage to reputation": Robert T. Hill, court-case documents, 1927, Papers of RTH.

197 "a combination of circumstances": Robert T. Hill, letter to Cornell University, December 1928, Papers of RTH.

198 "let the matter rest": Ralph Arnold, letter to Robert T. Hill, August 29, 1928, Papers of RTH.

198 "[Hill] seemed to think": Ralph Arnold, letter to C. A. Copper, June 9, 1928, Papers of RA.

198　had a lunch meeting: Ralph Arnold, diary, 1928, Papers of RA.

199　"this book," "this report": Robert T. Hill, court-case documents, 1927, Papers of RTH.

199　"Do not get the idea": Robert T. Hill, speech given to Building Owners and Managers Association, copy in Papers of HOW.

199　"Judging from past history": RTH 1928, 40.

199　"I don't believe": Robert Hill, unpublished document regarding alleged alterations in his book, Papers of RTH.

200　appears verbatim in a draft: Robert T. Hill, summary of conclusions, unpublished draft, Papers of RTH.

200　"If it should be technically proved": Robert Hill, draft of letter to editor, *Los Angeles Times*, undated, Papers of RTH.

200　"legal brief": William Morris Davis, "The Earthquake Problem in Southern California," galley proof, 1928, in Papers of RTH, box 34.

200　"book as finally printed," "Dr. Hill's claims of alteration": James M. Dixon, affidavit, June 1, 1928, Papers of RTH.

200　"technical hair-splitting": Charles Copper, letter to Lloyd Wright, January 31, 1928, Papers of RTH.

201　"the best years of [his] life," "hills and prairies [that] were," "These things," "The bitter later experiences": Robert T. Hill, letter to Justina Hill, May 1929, Papers of RTH.

201　"Part of my father's charm": Justina Hill in *FTG*, 272.

202　"A miracle has happened," "After a week," "a glorious day," "Saturday I return to Dallas": Robert T. Hill, letter to Justina Hill, May 1929, Papers of RTH.

202　He again reached out: Jesse Hill, letter to Robert T. Hill, October 26, 1932, Papers of RTH.

CHAPTER 10: RETRENCHMENT

204　"He will win": Sun Tzu, *The Art of War*, ca. 500 BC, translated by Peter Harris (New York: Everyman's Library/Random House, 2018).

204　$4,000 rebate, "with a view of reducing": "Earthquake Rate Slash Way Down," *Los Angeles Times*, May 2, 1929, A17.

205　"the recent earthquake at Calexico": Arthur Day, letter to Bailey Willis, March 1927, Papers of BW.

205　"may exceed $500,000": *Calexico Chronicle*, January 5, 1927, A1.

205　higher estimate of losses: "Fifty Shocks Wreck a Town in Mexico; California Shaken," *New York Times*, January 2, 1927, 1.

206 "A temblor": "Slight Temblor Felt," *Los Angeles Times*, July 9, 1929, A1.

206 "LOS ANGELES AND VICINITY IS VISITED BY EARTHQUAKE": *Trenton Evening News*, July 8, 1929, 1.

206 possible association between oil production and local earthquakes: S. E. Hough and M. Page, "Potentially Induced Earthquakes During the Early Twentieth Century in the Los Angeles Basin," *Bulletin of the Seismological Society of America* 106, no. 6 (2016): 2419–35.

207 "the most productive field," "the vacuum created": "Possible Causes of Earthquakes," *San Francisco Chronicle*, February 18, 1902, 6.

207 "These wiseacres," "What caused the earthquake": R. R., "All Quakes not Well Oiled," letter to the editor, *Los Angeles Times*, September 10, 1930, A4.

208 "Erroneous beliefs:" Ralph Arnold, "An Explanation of the California Earthquakes," undated, Papers of RA.

208 perturbing the stress: Paul Segall, "Stress and Subsidence Resulting from Subsurface Fluid Withdrawal in the Epicentral Region of the 1983 Coalinga Earthquake," *Journal of Geophysical Research* 90, no. B8 (1985): 6801–16; Susan E. Hough and Roger Bilham, "Poroelastic Stress Changes Associated with Primary Oil Production in the Los Angeles Basin, California," *Leading Edge* 37, no. 2 (2018): 108–16.

208 "I recognize that the information": Harry Wood, letter to Dr. H. W. Hoots, May 19, 1931, Papers of HOW.

209 "It is my belief that": Wayne Loel, letter to Edwin M. Daugherty, July 17, 1933, papers of RA.

210 "Regarding the matter of holding a meeting," "Anderson was the only": Harry Wood, letter to Stanford colleague, January 1928, Papers of HOW.

211 "no less than an extension": *FTG*, 261.

212 "hadn't seen that bad old frontier," "Oh, for a canteen": Robert T. Hill, "Unscrambling Misinterpreted Cabeza de Vaca and His Journey," *Dallas Morning News*, June 4, 1933.

212 "I would not exchange": Robert T. Hill, "Musings and Memories," *Dallas Morning News*, January 19, 1941.

212 "firm and unshakeable": *FTG*, 263.

212 "musing": Robert T. Hill, "Musings and Memories," *Dallas Morning News*, January 19, 1941.

213 "Willis' reputation as a seismologist": Sydney D. Townley, letter to Arthur Day, July 7, 1926, Papers of HOW.

214 "stir up the animals": Harry Wood, letter to Arthur Day, January 1928, Papers of HOW.

214 brought up the matter: Harry Wood, letters to Arthur Day, 1927, Papers of HOW.

215 "This man W.": Robert T. Hill, letter to Ellis Shuler, December 18, 1927, Papers of RTH.

215 "I have just read your paper": Ellis Shuler, letter to Robert T. Hill, January 14, 1928, Papers of RTH.

216 "Oddly enough, Day did not mention": Bailey Willis, letter to Margaret Willis, September 10, 1928, Papers of BW.

217 "safest place in any one": Bailey Willis, article about his address to the National Association of Building Owners and Managers, "Science Battles Menace of the Earthquake," *San Francisco Chronicle*, August 31, 1930, 11.

217 ongoing scientific investigations: Bailey Willis, "Man Probes the Secret of Earthquakes," *Baltimore Sun*, March 1931.

217 at least one of his books: Bailey Willis, *Living Africa: A Geologist's Wanderings Through the Rift Valley* (New York: Whittlesey House/McGraw-Hill, 1930), 320 pp.

217 articles on faults: Bailey Willis, "San Andreas Rift, California," *Journal of Geology* 46, no. 6 (1938): 793–827.

217 investigations in the Holy Land: B. Willis, "Dead Sea Problem: Rift Valley or Ramp Valley?" *Bulletin of the Geological Society of America* 39, no. 2 (1928): 490–542.

217 article on earthquakes: B. Willis, "Philippine Earthquakes and Structure," *Bulletin of the Seismological Society of America* 34, no. 2 (1944): 69–81.

217 short correction: Bailey Willis, "Earthquakes in the Holy Land: A Correction," *Nature* 131, no. 3311 (1933): 550.

218 earlier article: Bailey Willis, "Earthquakes in the Holy Land," *Bulletin of the Seismological Society of America* 18, no. 2 (1928): 73–103.

218 "It appears that . . . the statement": Arthur Day, letter to Bailey Willis, 1925, Papers of BW.

218 these long expeditions: Eliot Blackwelder, *Bailey Willis, May 31, 1857–February 19, 1949* (Washington, DC: National Academy of Sciences, 1961).

218 "Your wanderer must go": Bailey Willis, letter to Cornelia Grinnell Willis, October 2, 1902, Papers of BW.

219   "I look back": Bailey Willis, letter to Margaret Willis, September 20, 1940, Papers of BW.

219   "With the failure of Bailey Willis": CHG, 97.

219   "sound materials": ibid., 100.

CHAPTER 11: THE CLIMAX

222   "For it is not the light that is needed": Frederick Douglass, Fourth of July speech, 1852, in *The Life and Writings of Frederick Douglass*, vol. 2, *Pre–Civil War Decade 1850–1860*, edited by Philip S. Foner (New York: International Publishers, 1950).

222–23   "Nazis Unify Reich Grip," "Mussolini Hails Fascism's Spread," "Troops of Dictator in Vienna Now," "command of the United States fleet," "trod her historic decks": *Los Angeles Times*, March 11, 1933, 1.

223–24   Tony Gugliemo's younger sister: Selma Gugliemo, personal communication with author, 2016.

225   "many inches," "Proof that they did not slide": report to Harry Wood, Papers of HOW.

226   "I saw a two-story frame house": "Long Beach Counts Dead and Hurt After Quake Deals Mighty Blow," *San Bernardino Daily Sun*, March 11, 1933, 1.

226   tallied the list of dead: *Los Angeles Times*, March 12, 1933, 1.

226   "We tried to stand," "We suddenly realized": "Recalling the Long Beach Earthquake of 1933," *Orange County Register*, March 10, 2011.

226   "just once": "Last Fling at Chance Ends with His Life," *San Bernardino Daily Sun*, March 12, 1933, 2.

227   American professor at the University of Kobe in Japan: "Incidents During Earthquake," *Chicago Daily Tribune*, March 11, 1933, 3.

227   accidentally discharged his own rifle, died in a plane crash, "miscarried momentarily": ibid.

228   insurance liabilities were modest and manageable: "Insurance Survey Mapped," *Los Angeles Times*, March 12, 1933, 4.

228   Polytechnic High School: "Courageous Compton," *Riverside Daily Press*, June 2, 1933, 1.

229   "The rebuilding of Long Beach and Compton": "Clearing Up Quake Debris Gives Work to Nearly Two Thousand of City's Jobless," *Riverside Daily Press*, March 13, 1933, 2.

229 "AN OPPORTUNITY—AND A FAILURE," "Out of the Dust," "PLANS FOR IMMEDIATE": "An Opportunity—and a Failure," *Los Angeles Times*, March 12, 1933, 1.

229 "Whatever legislative relief": "Gears Greased by Legislature," *Los Angeles Times*, March 13, 1933, 8.

231–32 the Field Act, "An act relating to the safety": Robert A. Olson, "Legislative Politics and Seismic Safety: California's Early Years and the 'Field Act,' 1925–1933," *Earthquake Spectra* 19, no. 1 (2003): 111–31.

232 "they would blast them": ibid., 122.

233 adopted the building code: CHG, 112. "a masterful brief": ibid., 110.

233 pried loose funds from other sources: "Witness Lauds Newer Schools," *Los Angeles Times*, March 23, 1933, A1.

233 rate of investment in Los Angeles: "March Building Permits for Los Angeles and San Francisco Vicinities Classified," *Los Angeles Times*, April 16, 1933, 19.

233 rate of population growth: www.laalmanac.com/population.

CHAPTER 12: SETTLING THE SCORE

235 "We must not say": Marcus Tullius Cicero, *De Divinatione*, 44 BC, Act II, Scene 22.

235 "The Long Beach earthquake": Charles Richter, *Elementary Seismology* (San Francisco: W. H. Freeman and Co., 1958), 498.

236 "a prophet who would rather": Bailey Willis, letter to Harry Wood, March 14, 1933, Papers of HOW.

236 "only a modestly strong local shock": Harry Wood, letter to Bailey Willis, Papers of HOW.

236 "nevertheless not a general one": "More Quakes Are Expected," *Evansville Press*, March 22, 1933, 12.

236 "The Long Beach earthquake appears," "The disaster emphasizes": "A Prophecy in 1926 Warned California of Its Disaster," *Kansas City Star*, March 12, 1933, 7.

237 "is . . . free from a probability," "it cannot be said": RTH 1928.

238 "most photographed commercial": Earthquake '33, in *Historical Society of Long Beach Journal 1980–1981* (Long Beach, CA: Historical Society of Long Beach, 1933).

238　"The Long Beach earthquake": Bailey Willis, letter to John Merriam, March 31, 1933, Papers of BW.

238　a souvenir booklet, "The situation is in good hands," "Not a single Class A building," "The best scientific data": *Historical Society of Long Beach Journal 1980–1981*, ibid.

240　"There is no evidence": RTH 1928, 41.

241　"If you are on the fault": ibid., 37.

242　In the 2010 magnitude-7.0 Haiti earthquake: Susan E. Hough, Jean Robert Altidor, Dieuseul Anglade, Doug Given, M. Guillard Janvier, J. Zebulon Maharrey, Mark Meremonte, Bernard Saint-Louis Mildor, Claude Prepetit, and Alan Yong, "Localized Damage Caused by Topographic Amplification During the 2010 M7.0 Haiti Earthquake," *Nature Geoscience* 3, no. 11 (2010): 778.

242　"in some respects," "The banquet was beautiful," "Even the great lion": Robert T. Hill, letter to Justina Hill, April 9, 1933, Papers of RTH.

CHAPTER 13: THEATER

244　"All the world's a stage": William Shakespeare, *As You Like It*, 1623, Act II, Scene 7.

244　The initial article: Robert T. Hill, "Growing Pains of California's Mountains Cause Earthquakes, Noted Geologist's Explanation," *Dallas Morning News*, March 12, 1933, 8.

244–45　"Southern California has realized," "Neither need we fear": ibid.

245　"almost priceless pools of oil," "If the epicenter": "Los Angeles Finds Itself," *Dallas Morning News*, April 11, 1933, 8.

245　"As severe as": ibid.

247　"A reputation for seismicity": CHG, 89.

248　"very pronounced fault," "extensive disturbances": Ralph Arnold, speech to business leaders, New York City, July 17, 1920, Papers of RA.

248　"stated that he thought": Ford Carpenter, statement in minutes of the earthquake committee of the Los Angeles Chamber of Commerce, September 7, 1920, Papers of RA.

249　"It is strongly probable": Harry Wood, statement in response to request from Ford A. Carpenter, September 1927, Papers of HOW.

249　"To refuse to recognize": CHG.

250　"little quivers," "an innocent expression": "Hollywood and Earthquakes," *Los Angeles Times*, September 7, 1923, I14.

250 "For many years": Ralph Arnold, undated speech, audience unknown, ca. 1925–26, Papers of RA.

251 ask for maps or literature on the locations of faults: G. C. Ward, letter to Ralph Arnold, July 11, 1925, Papers of RA.

251 "It was also largely through": Chief Deputy of Public Works, letter to Harry Wood, late 1930, Papers of HOW.

252 a Shell Oil geologist: 1933, Papers of HOW.

252 "Architects, contractors, public service corporations": Ralph Arnold, undated speech, audience unknown, ca. 1925–26, Papers of RA.

252 garnered a banner headline: "SEVERE EARTHQUAKE WRECKS SANTA BARBARA; HOTELS COLLAPSE; BUSINESS AREA IS IN RUINS; 12 PERSONS KILLED; PROPERTY LOSS OF MILLIONS," *New York Times*, June 30, 1925, 1.

252 a later article: "LAYS QUAKE DAMAGE TO FAULTY BUILDINGS," *New York Times*, July 24, 1925, 25.

253 California Development Association: CHG, 73. provisions for earthquake-resistant design: ibid., 88.

253 the city of Santa Barbara formally requested: "Santa Barbara Asks for Advice on Rebuilding," *Los Angeles Times*, July 1, 1925, 2.

255 "That earthquake, all of the features of which," "Still you must bear in mind," "will make a good report": Arthur Day, letter to Harry Wood, January 13, 1923, Papers of HOW.

255 "one of the best scarers alive": Bailey Willis, letter to John Buwalda, April 6, 1940, Papers of BW.

255 "scare people into action," CHG, 84.

EPILOGUE: LEGACIES AND LESSONS

258 "Science, my lad": Jules Verne, *Journey to the Center of the Earth* (Paris: Pierre-Jules Hetzel, 1864).

258 "The main tool they used": CHG, 90.

259 "It was his practice," "Hill's sight was almost gone": *FTG*, 272.

259 "so deaf": Frederic H. Lahee, "Robert Thomas Hill, 1858–1941," *Science*, September 12, 1941, 249–50.

259 "Claude, this hotel": *FTG*, 271.

261 married his third wife, Sue Johnson: *FTG*, 275.

261 "beloved and faithful secretary": typed will, June 9, 1941, Papers of RTH.

262 "The Good Lord broke the pattern": Herbert Gambrell, letter to T. Wayland Vaughan, November 14, 1944, Papers of T. Wayland Vaughan, Smithsonian Archives, Washington DC.

262 "Dr. Hill came more nearly": Charles N. Gould, "The Passing of a Great Geologist—Robert T. Hill," in *Proceedings of the Oklahoma Academy of Science* 22 (2015): 150–51.

262 "The old gentleman faced west": *FTG*, 270.

262 "You see more": Charles Gould, letter to Robert T. Hill, February 26, 1930, Papers of RTH.

263 "Wegenerism," "a passing squall": Bailey Willis, letter to Arthur Day, June 4, 1927, Papers of BW.

264 "almost the last of that galaxy": Charles Gould, "The Passing of a Great Geologist—Robert T. Hill," in *Proceedings of the Oklahoma Academy of Science* 22 (2015): 150–51.

266 "choose men with great care": Ralph Arnold, letter to acting president of Caltech, November 1920, Papers of RA.

267 "The doctors": Robin Willis, letter to Margaret Willis, March 19, 1941, Papers of BW.

267 Margaret died: *San Francisco Chronicle*, March 23, 1941.

267 "I am as busy as ever": Bailey Willis, letter to Caroline Greeley Carpenter, July 10, 1947, Papers of BW.

267 "Dear Grand Young Man": Ralph Arnold, letter to Bailey Willis, May 29, 1947, Papers of RA.

267 "His vigorous bearing": "Earthquake Willis," *Newsweek* magazine, June 2, 1947.

269 "When I received your letter," "whilst researching," "struck me as sort of": Petrie Mondell, letter to Bailey Willis, February 12, 1934, Papers of BW.

269 "Thanks for your comment," "The book itself": Bailey Willis, letter to Petrie Mondell, February 16, 1934, Papers of BW.

269 "man who never accepted his own worth": *FTG*, 271.

270 "common human error": Petrie Mondell, letter to Bailey Willis, February 12, 1934, Papers of BW.

272 "I maintain": Robert T. Hill, "Current Erroneous Impressions Concerning Earthquake Effects," undated speech, Papers of RTH.

SELECTED BIBLIOGRAPHY

Alexander, Nancy. *Father of Texas Geology*. Dallas: Southern Methodist University Press, 1976. 317 pp. (Abbreviated *FTG* in notes)

Anderson, J. A., R. Arnold, W. W. Campbell, A. L. Day (chairman), A. C. Lawson, R. A. Millikan, H. F. Reid, and B. Willis. "Report of Advisory Committee on Seismology of the Carnegie Institution of Washington." *Bulletin of the Seismological Society of America* 13, no. 4 (1922): 231–37.

Biasi, Glenn P., and Katherine M. Scharer. "The Current Unlikely Earthquake Hiatus at California's Transform Boundary Paleoseismic Sites." *Seismological Research Letters* 90, no. 3 (2019): 1168–76.

Biasi, Glenn P., and Ray J. Weldon. "San Andreas Fault Rupture Scenarios from Multiple Paleoseismic Records: Stringing Pearls." *Bulletin of the Seismological Society of America* 99, no. 2A (2009): 471–98.

Boles, J. R., G. Garven, H. Camacho, and J. E. Lupton. "Mantle Helium along the Newport-Inglewood Fault Zone, Los Angeles Basin, California: A Leaking Paleo-subduction Zone." *Geochemistry, Geophysics, Geosystems* 16, no. 7 (2015): 2364–81.

Chipman, Donald E. "In Search of Cabeza de Vaca's Route across Texas: A Historiographical Survey." *Southwestern Quarterly* 91 (1987): 175–86.

Field, E. H., G. P. Biasi, P. Bird, T. E. Dawson, K. R. Felzer, D. D. Jackson, K. M. Johnson, T. H. Jordan, C. Madden, A. J. Michael, K. R. Milner, M. T. Page, T. Parsons, P. M. Powers, B. E. Shaw, W. R. Thatcher, R. J. Weldon II, and Y. Zeng. *Uniform California Earthquake Rupture Forecast, Version 3 (UCERF3)—The Time-Independent Model*. USGS Open-File Report 2013-1165, CGS Special Report 228, and Southern California Earthquake Center Publication 1792. Washington, DC: US Department of the Interior, 2013. 97 pp. http://pubs.usgs.gov/of/2013/1165.

Geschwind, Carl-Henry. *California Earthquakes: Science, Risk, and the Politics of Risk Mitigation.* Baltimore: John Hopkins University Press, 2001. 337 pp. (Abbreviated CHG in notes)

Goodwin, Judith R. *Millikan's School: A History of the California Institute of Technology.* New York: W. W. Norton, 1991. 317 pp.

Hill, Robert Thomas. "Do We Teach Geology?" *Popular Science Monthly* 40 (November 1891).

———. "A Study of Pelee: Impressions and Conclusions of a Trip to Martinique." *Century Illustrated Monthly* 64 (May–October 1902): 764–85.

———. "Earthquake Conditions in Southern California." In the proceedings of the Fortieth Annual Meeting of the Geological Society of America, Cleveland, Ohio, December 29–31, 1927, printed in *Bulletin of the Geological Society of America* 39 (1928): 188–89.

———. *Southern California Geology and Los Angeles Earthquakes.* Los Angeles: Arts Printing Company, copyright C. A. Copper, 1928. 232 pp. (Abbreviated RTH 1928 in notes)

Hough, Susan E. *Richter's Scale: Measure of an Earthquake, Measure of a Man.* Princeton, NJ: Princeton University Press, 2007. 335 pp.

Page, Morgan, and Karen Felzer. "Southern San Andreas Fault Seismicity Is Consistent with the Gutenberg–Richter Magnitude–Frequency Distribution." *Bulletin of the Seismological Society of America* 105, no. 4 (2015): 2070–80.

Scharer, Katherine, Ray Weldon, Ashley Streig, and Thomas Fumal. "Paleo-earthquakes at Frazier Mountain, California Delimit Extent and Frequency of Past San Andreas Fault Ruptures along 1857 Trace." *Geophysical Research Letters* 41, no. 13 (2014): 4527–34.

Steiny, H. M. "Memorial: Ralph Arnold." *Bulletin of the American Association of Petroleum Geologists* 45, no. 11 (November 1961): 1897–906.

Townley, S. D. "John Casper Branner (Memorial)." *Bulletin of the Seismological Society of America* 12, no. 1 (1922): 1–11.

Willis, Bailey. *A Yanqui in Patagonia: A Bit of Autobiography.* Stanford, CA: Stanford University Press, 1947. 152 pp. (Abbreviated *AYP* in notes)

INDEX

Italic page numbers indicate illustrations.

Shuler, Ellis, 72, 150–52, 171–72, 201–2, 215, 260

Siberia, 24

Sierra Nevada, 4, 14

Smith, Skookum, 27

Smith College, 106

Smithsonian Institution, 52–53

South America, 99, 190

South Cherry Street, 34, 36

Southern California Academy of Sciences, 178–79, 192, 194

Southern California Branch of the Seismological Society of America, 154

Southern California Council for Earthquake Protection, 160, 168, 249

Southern California Edison, 251, 253

Southern California Geology and Los Angeles Earthquakes (Hill), 186–92; aftermath of publication of, 202–3, 204–6, 211–13, 256–57; dust cover of, 186; legal case over, 197–201; plans for, 177, 178–81; publication process, 185; publicity about, 181–83, 194–96; reviews of, 192–94; statements viewed in hindsight after Long Beach earthquake, 235–42

Southern Democrat. *See* Hill, Robert Thomas, Jr.: political affiliation of

Southern Methodist University, 72, 170–71, 201, 259

South Pacific, 190

Spain, 150

Standard Oil Company, 251

Stanford University, 117, 120, 123–24, 128, 134, 139, 210, 213; Bailey Willis career at, 99–100, 103–4, 136, 169–70, 242, 267–68; geology department, founding of, 77, 103–4

State Street, 141–42

St. Francis Dam, collapse of (1928), 183–85, 194, 211, 232, 234, 245, 272

St. John, Adela Rogers, 108

St. Louis, 44

Strait of Messina earthquake (1909), 110, 179

Structural Engineers Association of Southern California, 254

Stuttgart (Germany), 22

Summerland oil field, 140

Taber, Stephen, 128–29

Tacoma (Washington), 26–27

Tarr, Ralph, 62–63

Taylor, Thomas Ulvin, 59–60

Temblor Range, 5

Tennessee, 41–44, 57, 76–77, 90, 96, 110, 150; Civil War battles in, 33–34; Hospital for the Insane, 37

Texas: Supreme Court, 150. *See also* Hill, Robert Thomas, Jr.; *and individual cities*

Tokyo earthquake (1923), 220, 250

Towne, Charles, 227

Townley, Sidney, 213–14

Trans-Pecos region, 82, 84

Trans-Pecos report, 93–94, 96, 101

Treat, Payson, 170

Twain, Mark, 47

Uniform Building Code, 253

Union Oil Company, 208, 251

United States Coast and Geodetic Survey, 133, 145, 160–61, 176–77, 180–82, 188, 218; earthquake monitoring program of, 221, 265, 266

United States Geological Survey: Appalachian Division of, 30; contributions of, 51; earthquake studies by, 91–92; ethics rules of, 95; founding of, 23, 102, 120, 265; mapping activities of, 65–66, 139; prediction by Bailey Willis, comment on, 180–82, 188. *See also individuals' work at US Geological Survey*

United States Navy, 208

United States Quartermaster, 27, 39

United States Supreme Court, 77

University of California, Berkeley, 103, 117–19, 123, 132
University of California, Los Angeles (UCLA), 174, 213, 242
University of Texas, Austin, 57, 69, 152
unreinforced masonry construction, 164; vulnerability of, 141–42, 224, 228, 239, 252

Valencia (California), 184
Vanderbilt University, 45–46
Vaughan, T. Wayland, 171–72
Ventura, 158
Veracruz earthquake (1920), 127
Vermilion Iron Range, 25
Virginia, 77, 156

Waco (Texas), 41, 45
Wade, Ruby, 227
Walcott, Charles, 65, 71, 82, 94–95
Washington, DC: Bailey Willis in, 29, 79; Harry Wood in, 118, 121, 130; Ralph Arnold in, 124, 157; Robert T. Hill in, 55, 64, 76, 92–93, 110, 180–81, 201, 215
Washington State, 24, 26
Washington Times, 58
Wegener, Alfred, 112, 263
West Indies, 85, 171
White, Dr. Charles Abiathar, 52–57, 78, 201
Whittier earthquake (1929), 205–6; relationship to oil production, 206–7
Whittier Narrows earthquake (1987), 162
Wiesendanger, Emil Ulrich "Wies," 45–46
Wilkins curse, 108–9
Willis, Bailey, 15, *19*, 125, 242–43; articles by, 8, 10–11, 217–18; California, fieldwork in, 102; California, move to, 99–101; career at US Geological Survey, 23–24, 29–31, 50–52; childhood, 16–22, 32, 34; committee activities of, 133; death, 268; education, 21–23, 26;

expeditions, 98–99, 173–74; family, 138–39, 218–19; Iron Range, work in, 25–26; later years, 267–68; legacy, 265–66, 266–67; Long Beach earthquake, activities after, 236–37, 242–44; marriages, 28, 99; Pacific Northwest, work in, 26–29; personality, 169–73; physical appearance, 32–33; plate tectonics, views of, 263; political views, 104, 116, 173, 257; prediction, fallout from, 213–17; prediction by, 153, 160–66, 168–69, 176, 180–81, 191, 252–53, 256, 269–70; Robert T. Hill, interactions with, 31; Robert T. Hill, views of, 269–70; Santa Barbara earthquake, experience with, 139–46; scientific contributions of, 264–65; Southerners, views of, 70–71, 76–78; Stanford, career at, 103–4, 267–68; St. Francis Dam collapse, investigation of, 184–85; university education, 22–23; US Geological Survey, early work with, 23–24; US Geological Survey, later work with, 29–31, 50–52, 71–74; "Windy Willis" moniker, 170–73, 255, 262, 267
Willis, Cornelia "Nellie" Grinnell, 17–18, 20–23, 25–26, 66–67, 218
Willis, Cornelius "Neal," 99, 139
Willis, Edith, 17, 21, 99, 138
Willis, Grinnell "Nelt," 21
Willis, Hope, 29, 76, 99, 138
Willis, Imogene, 17
Willis, Margaret (daughter of Bailey Willis), 100, 170
Willis (née Baker), Margaret (wife of Bailey Willis), 99, 138, 173, 174, 217, 267
Willis, Nathaniel Parker, 17–18, 20–21, 68
Willis, Robin, "Bob," 99, 136, 139
Wilmington oil field, 208
Wilson High School, 223–24
Windward Islands, 96
Wood, Harry Oscar, 117, *119*, 142, 154, 199, 205–6, 208, 210–11, 247–52, 266;